Problem-Solving Guide with Solutions
to accompany Philip R. Kesten and David L. Tauck's

UNIVERSITY PHYSICS

for the Physical and Life Sciences   *Volume I*

Timothy A. French
DePaul University

W. H. FREEMAN AND COMPANY
New York

© 2013 by W. H. Freeman and Company

Printed in the United States of America

ISBN-13: 978-1-4641-0096-3
ISBN-10: 1-4641-0096-9

First printing

W. H. Freeman and Company
41 Madison Avenue
New York, NY 10010
RG21 6XS England

www.whfreeman.com/physics

Contents

About the Author

DR. TIMOTHY FRENCH

Dr. Timothy French is currently a Visiting Assistant Professor in the Department of Chemistry at DePaul University in Chicago, IL. Dr. French holds a BS in chemistry from Rensselaer Polytechnic Institute and a MS and PhD in chemistry from Yale University. His undergraduate research looked at how to assess problem-solving ability in introductory physics students, while his doctoral research was on tera-hertz spectroscopy. Prior to arriving at DePaul, he was a Preceptor in Chemistry and Chemical Biology for four years at Harvard University where he taught an introductory physics sequence with biological and medical applications, experimental physical chemistry, and introductory physical chemistry, as well as the introductory physics class at the Harvard Summer School. He was awarded the Harold T. White Prize for Excellence in Teaching from the Harvard Physics Department in 2009.

A Note from the Author

The Problem-Solving Guide with Solutions is more than a book of answers to physics problems. It's designed to help you learn the way physicists approach and solve problems.

Many students, when approaching a physics problem, will half-read the question, pick out some of the variables used in the problem statement, and then start blindly hunting for an equation that shares the same letters, regardless of its utility.

Physicists, on the other hand, approach problems very differently. They'll fully read the question, underlining important information as they go. Next they will draw their own picture, even if one is provided. Then they will consider the underlying concept at play ("Is this a conservation of momentum question? Is Newton's second law best used here?") before determining which equation will be the most helpful. The equations are all manipulated algebraically before putting in numbers and arriving at a numerical answer.

Students usually move on to the next problem at this point without performing the crucial final step—checking the answer for reasonability. Physicists make sure the answer has the correct units, check some limiting cases ("What happens if the angle goes to 0 degrees? 90 degrees?"), and perform order of magnitude estimations all in order to gain confidence in their final answer.

Each solution in this book is set up to mirror this process with three distinct parts: **Set Up, Solve,** and **Reflect**. The **Set Up** portion contains all of the logic behind starting the problem, from a rehash of the important information in the problem statement to the conceptual underpinnings necessary in understanding the solution. The Solve portion contains the algebraic steps used to arrive at the numerical solution; note that this step comes *after* the Set Up step. Finally, the Reflect step provides a "sanity check" for the answer—"Is this what we expected? Does it make sense with respect to our observations of and interactions with everyday life? Does this answer seem reasonable?"

Very often instructors expect students to become expert problem solvers on their own without instruction. By explicitly showing the steps physicists use when solving problems, I hope you will be able to mirror and internalize these steps and make your foray into physics a more enjoyable and successful experience.

—Tim

Get Help with Premium Multimedia Resources

One of the benefits technology brings us in education is the ability to visualize concepts, gain problem-support, and test our skills outside of the traditional pen-and-paper classroom method. With that in mind, W. H. Freeman has developed a series of media assets geared at reinforcing conceptual understanding and building problem-solving skills.

P'Casts are videos that emulate the face-to-face experience of watching an instructor work a problem. Using a virtual whiteboard, the P'Casts' tutors demonstrate the steps involved in solving key worked examples, while explaining concepts along the way. The worked examples were chosen with the input of physics students and instructors across the country. P'Casts can be viewed online or downloaded to portable media devices.

Interactive Exercises are active learning, problem-solving activities. Each Interactive Exercise consists of a parent problem accompanied by a Socratic-dialog "help" sequence designed to encourage critical thinking as users do a guided conceptual analysis before attempting the mathematics. Immediate feedback for both correct and incorrect responses is provided through each problem-solving step.

Picture Its help bring static figures from the text to life. By manipulating variables within each animated figure students visualize a variety of physics concepts. Approximately 50 activities are available.

These Premium Multimedia Resources are available to you in a few places:

- If you're using the printed text, you can purchase the Premium Media Resources for a small fee via the Book Companion Website (BCS) at www.whfreeman.com/universityphysics1e.
- If you've purchased the W. H. Freeman media-enhanced eBook, these resources are embedded directly into the chapters of the text and available on the BCS.
- If you're using an online homework system, such as WebAssign or PhysicsPortal, these resources are integrated into your individual assignments and are available on the BCS.

And while all three of those places correlate the Premium Media Resources by chapter, we've gone one step further in the Problem-Solving Guide with Solutions and correlated each media resource by problem. Look for each relevant item called out after its corresponding problem:

> **Get Help:** Picture It - Adding and Subtracting Vectors

To get started using these exciting resources, log on to www.whfreeman.com/universityphysics1e!

Chapter 1
Physics: An Introduction

Conceptual Questions

1.3 Use of the metric prefixes makes any numerical calculations much easier to follow. Instead of an obscure conversion (12 in/ft, 1760 yards/mi, 5280 ft/mi), simple powers of 10 make the transformations (10 mm/cm, 1000 m/km, 10^{-6} m/μm).

1.7 No. The equation "3 meters = 70 meters" has consistent units but it is false. The same goes for "1 = 2," which consistently has no units.

1.9 The SI unit for length is the meter; the SI unit for time is the second. Therefore, the SI units for acceleration is meters/(second)^2, or m/s^2.

Multiple-Choice Questions

1.13 C (10^4).

$$1 \text{ m}^2 \times \frac{100 \text{ cm}}{1 \text{ m}} \times \frac{100 \text{ cm}}{1 \text{ m}} = \boxed{10^4 \text{ cm}^2}$$

1.17 C (1810). Both 25.8 and 70.0 have three significant figures. When multiplying quantities, the quantity with the fewest significant figures dictates the number of significant figures in the final answer. Multiplying 25.8 by 70.0 gives 1806, which has four significant figures. Our final answer must have three significant figures, so we round 1806 to 1810.

1.19 B (have dimensions of 1/T). An exponent must be dimensionless, so the product of λ and t must be dimensionless. The dimension of t is T. Therefore, λ has dimensions of 1/T.

Estimation Questions

1.21 We can model Mt. Everest as a 45° triangular pyramid—three identical triangles angled at 45° from an equilateral triangle base. The volume of a triangular pyramid is $\frac{1}{3}$ (area of base)(height of pyramid). The base of Mt. Everest is 4500 m above sea level, and its peak is 8800 m above sea level, so its height is 4300 m. We can calculate (from geometry) the length of each side of the base equilateral triangle; each side is $\frac{12,900}{\sqrt{3}}$. Therefore the volume of Mt. Everest is approximately

$$\frac{1}{3}\left(\frac{1}{2}\right)\left(\frac{12900}{\sqrt{3}} \text{ m}\right)(6450 \text{ m})(4300 \text{ m}) \approx 3 \times 10^{10} \text{ m}^3.$$ The density of rock is about

2750 kg/m³. The mass of Mt. Everest is then $m_{Everest} = \rho_{rock} V_{Everest} =$

$$(3 \times 10^{10} \text{ m}^3) \times \left(\frac{2750 \text{ kg}}{\text{m}^3}\right) \approx \boxed{10^{14} \text{ kg}}.$$

1.25 We can split an average student's daily water use into four categories: showering, cooking/drinking/hand-washing, flushing the toilet, and doing laundry. A person uses about 100 L of water when showering, about 10 L for cooking/drinking/hand-washing, about 24 L when flushing the toilet, and about 40 L when doing two loads of laundry. This works out to about 150–200 L of water per day.

Get Help: P'Cast 1.1 – Hair Growth

1.29 We can estimate the number of cells in the human body by determining the mass of a cell and comparing it to the mass of a human. An average human male has a mass of 80 kg. A person is mainly water, so we can approximate the density of a human body (and its cells) as 1000 kg/m³. We are told that the volume of a cell is the same as a sphere with a radius of 10^{-5} m, or approximately 4×10^{-15} m³; the mass of a single cell is $4 \times 10^{-15} \text{ m}^3 \times \dfrac{1000 \text{ kg}}{1 \text{ m}^3} = 4 \times 10^{-12}$ kg. The number of cells in the body is then

$$n_{cell} = \frac{m_{body}}{m_{cell}} = \frac{80 \text{ kg}}{4 \times 10^{-12} \text{ kg}} = \boxed{2 \times 10^{13}}.$$

Get Help: P'Cast 1.1 – Hair Growth

Problems

1.33

SET UP

A list of powers of 10 are given. We can use Table 1-3 in the text to determine the correct metric prefix associated with each factor. Eventually, knowing some of the more common prefixes will become second nature.

SOLVE

A) kilo (k)   E) milli (m)

B) giga (G)   F) pico (p)

C) mega (M)   G) micro (μ)

D) tera (T)   H) nano (n)

REFLECT

Whether or not the prefix is capitalized is important. *Mega-* and *milli-* both use the letter "m", but *mega-* is "M" and *milli-* is "m". Confusing these two will introduce an error of 10^9!

1.37

SET UP

This problem provides practice with metric unit conversions. We need to look up (and/or memorize) various conversions between SI units and hectares and liters. One hectare is equal to 10^4 m², and 1000 L is equal to 1 m³. Another useful conversion is that 1 mL equals 1 cm³.

SOLVE

A) $328 \text{ cm}^3 \times \dfrac{1 \text{ mL}}{1 \text{ cm}^3} \times \dfrac{1 \text{ L}}{1000 \text{ mL}} = \boxed{0.328 \text{ L}}$.

B) $112 \text{ L} \times \dfrac{1 \text{ m}^3}{1000 \text{ L}} = \boxed{0.112 \text{ m}^3}$.

C) $220 \text{ hectares} \times \dfrac{10^4 \text{ m}^2}{1 \text{ hectare}} = \boxed{2.2 \times 10^6 \text{ m}^2}$.

D) $44300 \text{ m}^2 \times \dfrac{1 \text{ hectare}}{10^4 \text{ m}^2} = \boxed{4.43 \text{ hectares}}$.

E) $225 \text{ L} \times \dfrac{1 \text{ m}^3}{1000 \text{ L}} = \boxed{0.225 \text{ m}^3}$.

F) $17.2 \text{ hectare} \cdot \text{m} \times \dfrac{10^4 \text{ m}^2}{1 \text{ hectare}} \times \dfrac{10^3 \text{ L}}{1 \text{ m}^3} = \boxed{1.72 \times 10^8 \text{ L}}$.

G) $2.253 \times 10^5 \text{ L} \times \dfrac{1 \text{ m}^3}{10^3 \text{ L}} \times \dfrac{1 \text{ hectare}}{10^4 \text{ m}^2} = \boxed{2.253 \times 10^{-2} \text{ hectare} \cdot \text{m}}$.

H) $2000 \text{ m}^3 \times \dfrac{1000 \text{ L}}{1 \text{ m}^3} \times \dfrac{1000 \text{ mL}}{1 \text{ L}} = \boxed{2 \times 10^9 \text{ mL}}$.

REFLECT
Rewriting the measurements and conversions in scientific notation make the calculations simpler and helps give some physical intuition.

1.43

SET UP
In the United States, fuel efficiency is reported in miles per gallon (or mpg). We are given fuel efficiency in kilometers per kilogram of fuel. We can use the conversions listed in the problem to convert this into mpg.

SOLVE

$$\dfrac{7.6 \text{ km}}{\text{kg}} \times \dfrac{0.729 \text{ kg}}{1 \text{ L}} \times \dfrac{3.785 \text{ L}}{1 \text{ gal}} \times \dfrac{1 \text{ mi}}{1.609 \text{ km}} = \boxed{13 \dfrac{\text{mi}}{\text{gal}}} \text{ or } \boxed{13 \text{ mpg}}.$$

REFLECT
This would be the gas mileage for a cargo van or large SUV. A hybrid sedan would have a gas mileage of around 40 mpg.

Get Help: P'Cast 1.1 – Hair Growth

1.49

SET UP
We are asked to calculate sums and differences with the correct number of significant figures. When adding or subtracting quantities, that the quantity with the fewest *decimal places* (not

significant figures) dictates the number of decimal places in the final answer. If necessary, we will need to round our answer to the correct number of decimal places.

SOLVE

A) $4.55 + 21.6 = \boxed{26.2}$ C) $71.1 + 3.70 = \boxed{74.8}$

B) $80.00 - 112.3 = \boxed{-32.3}$ D) $200 + 33.7 = \boxed{200}$

REFLECT

The answer to part D may seem weird, but 200 only has one significant figure—the "2". The hundreds place is, therefore, the smallest decimal place we are allowed.

1.53

SET UP

A string of length S is stretched into a rectangle with width W and length L. We need to find the ratio of W to L that yields the largest area. The area of the rectangle is $A = LW$. There is a constraint on the system: The perimeter of the rectangle, $(2L + 2W)$, must equal S. This lets us eliminate one of the variables, say, L. To find the value of W that maximizes the area A, we need to set the derivative of A with respect to W equal to zero and solve for W_{max}. Once we have W_{max}, we can solve for L_{max} and find the ratio between the two.

SOLVE

$$2L + 2W = S, \text{ so } L = \frac{S}{2} - W$$

$$A = LW = \left(\frac{S}{2} - W\right)W = \frac{S}{2}W - W^2$$

Maximizing the area:

$$\left(\frac{dA}{dW}\right)_{W_{max}} = 0 = \frac{S}{2} - 2W_{max}$$

$$W_{max} = \frac{S}{4}$$

Finding L_{max}:

$$2L_{max} + 2\left(\frac{S}{4}\right) = S$$

$$L_{max} = \frac{S}{4}$$

$$\frac{W_{max}}{L_{max}} = \frac{\left(\dfrac{S}{4}\right)}{\left(\dfrac{S}{4}\right)} = \boxed{1}$$

REFLECT

A square is a rectangle where $L = W$. A square is known to maximize the area for a fixed perimeter, which is what we explicitly calculated.

1.57

SET UP

We are given an equation that describes the motion of an object and asked if it is dimensionally consistent; we need to make sure the dimensions on the left side equal the dimensions on the right side. The equation contains terms related to position, speed, and time. Position has dimensions of length; speed has dimensions of length per time; and time has dimensions of, well, time.

SOLVE

$$x = vt + x_0$$

$$[L] \stackrel{?}{=} \frac{[L]}{[T]}[T] + [L]$$

$$[L] = [L] + [L]$$

REFLECT

Dimensions are general (for example, length), while units are specific (for example, meters, inches, miles, furlongs).

1.61

SET UP

We need to check that the dimensions of the expected quantity match the dimensions of the units reported. Remember that dimensions should be given in terms of the fundamental quantity (see Table 1-1 in the text). For example, volume has dimensions of (length)^3.

SOLVE

A) Volume flow rate has dimensions of volume per time, or $\frac{[L]^3}{[T]}$, so $\frac{m^3}{s}$ is $\boxed{\text{correct}}$.

B) Height has dimensions of $[L]$. Units of m^2 is $\boxed{\text{not correct}}$.

C) A fortnight has dimensions of $[T]$, while m/s are the units of speed. This statement is $\boxed{\text{not correct}}$.

D) Speed has dimensions of $\frac{[L]}{[T]}$, but $\frac{m}{s^2}$ are units of acceleration. This statement is $\boxed{\text{not correct}}$.

E) Weight is a force and has dimensions of $\frac{[M][L]}{[T]^2}$, and lb is an appropriate unit for force. This is $\boxed{\text{correct}}$.

F) Density has dimensions of mass per volume, or $\frac{[M]}{[L]^3}$, so $\frac{kg}{m^2}$ is $\boxed{\text{not correct}}$.

REFLECT

Only statements A and E are correct. Making sure your answer has the correction dimensions and units is an important last step in solving every problem.

1.67

SET UP

A 2-m-tall woman stands in front of a flagpole of height h. The top of the woman's shadows and the top of the flagpole's shadow overlap. The woman's shadow is 10 m long, while the flagpole's shadow is 22 m long. The woman and the flagpole each create a triangle with their shadows. These triangles are similar to one another. Therefore, we know that the ratio of the woman's height to the size of her shadow is equal to the ratio of the flagpole's height h and the size of its shadow.

Figure 1-1 Problem 67

SOLVE

$$\frac{2 \text{ m}}{10 \text{ m}} = \frac{h}{22 \text{ m}}$$

$$h = \left(\frac{2 \text{ m}}{10 \text{ m}}\right)(22 \text{ m}) = \boxed{4.4 \text{ m}}$$

REFLECT

This works out to a little over 14 feet, which is a reasonable height for a flagpole.

1.71

SET UP

It takes about 1 min for all of the blood in a person's body to circulate through the heart. When the heart beats at a rate of 75 beats per minute, it pumps about 70 mL of blood per beat. Blood has a density of 1060 kg/m³. We can use each of these relationships as a unit conversion to determine the total volume of blood in the body and the mass of blood pumped per beat. For the volume calculation, we will start with the fact that all of the blood takes 1 min. to circulate through the body. Because we are interested in the mass per beat, we can start with 1 beat in part b.

SOLVE

Part a)

$$1 \text{ min} \times \frac{75 \text{ beats}}{1 \text{ min}} \times \frac{70 \text{ mL}}{1 \text{ beat}} \times \frac{1 \text{ L}}{1000 \text{ mL}} = \boxed{5.25 \text{ L}} \times \frac{1 \text{ m}^3}{1000 \text{ L}} = \boxed{5.25 \times 10^{-3} \text{ m}^3}.$$

Part b)

$$1 \text{ beat} \times \frac{70 \text{ mL}}{1 \text{ beat}} \times \frac{1 \text{ L}}{1000 \text{ mL}} \times \frac{1 \text{ m}^3}{1000 \text{ L}} \times \frac{1060 \text{ kg}}{1 \text{ m}^3} = \boxed{0.074 \text{ kg}} \times \frac{1000 \text{ g}}{1 \text{ kg}} = \boxed{74 \text{ g}}.$$

REFLECT

A volume of 5.25 L is about 11 pints, which is about average. For comparison, they take 1 pint when you donate blood.

Get Help: P'Cast 1.1 – Hair Growth

Chapter 2
Linear Motion

Conceptual Questions

2.1 An object will slow down when its acceleration vector points in the opposite direction to its velocity vector. Recall that acceleration is the change in velocity over the change in time.

2.5 Speed and velocity have the same SI units (m/s). Speed is the magnitude of the velocity vector. If the velocity points completely in the positive direction, then the two can be interchanged. If the velocity is not fixed in direction, then the correctly signed component of the velocity will need to be used to avoid confusion.

2.9 The average velocity of a moving object will be the same as the instantaneous velocity if the object is moving at a constant velocity (both magnitude and direction). Also, if the average velocity is taken over a period of constant acceleration, the instantaneous velocity will match it for one moment in the middle of that period.

2.13 Assuming the initial speed of the ball is the same in both cases, the velocity of the second ball will be the same as the velocity of the first ball. The upward trajectory of the first ball involves the ball going up and reversing itself back toward its initial location. At that point, the trajectories of the two balls are identical as the balls hit the ground.

2.19 Part a) If the police car starts at rest and eventually catches up to a car traveling at a constant speed, the police car must be traveling faster than the speeding car. Therefore, at the point the police car overtakes the speeding car, the police car's speed is greater than that of the speeding car.

Part b) The displacement of the police car is the same as the displacement of the speeding car. If both cars start at the location of the police car, travel along the same path, and end at the same location, the displacements of the two cars will be the same.

Part c) The acceleration of the police car will be greater than the acceleration of the speeding car. The police car starts at rest and must accelerate up to some final speed in order to catch up with the speeding car. The speeding car is traveling at a constant speed, so its acceleration is zero, assuming it's traveling in a straight line.

Multiple-Choice Questions

2.23 D.

The slope of a position versus time plot gives information regarding the speed of the object. Point D has the largest slope, which means the object is moving the fastest there.

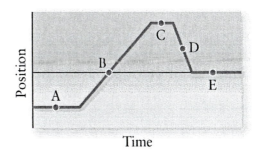

Figure 2-1 Problem 23

2.27 **A** (increasing). If the car's velocity and acceleration vectors point in the same direction, the car will speed up. Because the magnitude of the acceleration is decreasing with time (but always pointing in the same direction), the *rate at which the car speeds up* will decrease.

Estimation Questions

2.33 A swimmer completes a 50-m lap in 100 s. His average speed is equal to

$$v_{average} = \frac{50 \text{ m}}{100 \text{ s}} \times \frac{3600 \text{ s}}{1 \text{ hr}} \times \frac{1 \text{ km}}{1000 \text{ m}} = \boxed{1.8 \frac{\text{km}}{\text{hr}}}$$

Get Help: P'Cast 2.1 – Down the Stretch

2.35 The fastest time for the 100-m race for men was set by Usain Bolt (9.58 s) and for women by Florence Griffith-Joyner (10.49 s). These times correspond to top speeds of 10.4 m/s and 9.5 m/s, respectively. A runner who falls in the mud will take much longer to come to rest than one who falls on a running track. Let's assume it takes 2 s to come to rest in the mud but only 0.5 s on a track. The magnitudes of the acceleration for each case are

$$a_{\text{Bolt, mud}} = \frac{\Delta v}{\Delta t} = \frac{10.4 \frac{\text{m}}{\text{s}}}{2 \text{ s}} = \boxed{5.2 \frac{\text{m}}{\text{s}^2}}$$

$$a_{\text{FloJo, mud}} = \frac{\Delta v}{\Delta t} = \frac{9.5 \frac{\text{m}}{\text{s}}}{2 \text{ s}} = \boxed{4.8 \frac{\text{m}}{\text{s}^2}}$$

$$a_{\text{Bolt, track}} = \frac{\Delta v}{\Delta t} = \frac{10.4 \frac{\text{m}}{\text{s}}}{0.5 \text{ s}} = \boxed{21 \frac{\text{m}}{\text{s}^2}}$$

$$a_{\text{FloJo, mud}} = \frac{\Delta v}{\Delta t} = \frac{9.5 \frac{\text{m}}{\text{s}}}{0.5 \text{ s}} = \boxed{19 \frac{\text{m}}{\text{s}^2}}$$

2.37 On the open sea, a cruise ship travels at a speed of approximately 10 m/s. We need to estimate the time it takes the ship to reach its cruising speed from rest. We should expect that it takes more than a few minutes but less than a full hour; let's estimate the time to be 0.5 hr. The magnitude of the average acceleration of the cruise ship is the change in the ship's speed divided by the time interval over which that change occurs:

$$a_{average} = \frac{\Delta v}{\Delta t} = \frac{10 \frac{\text{m}}{\text{s}}}{0.5 \text{ hr}} \times \frac{1 \text{ hr}}{3600 \text{ s}} = \boxed{0.006 \frac{\text{m}}{\text{s}^2}}$$

2.41 We can make a plot of position versus time and determine the equations for each region by manually calculating the slope and *y* intercepts or by using a computer program to fit the data in each time interval.

$t(\mathrm{s})$	$x(\mathrm{m})$
0	-12
1	-6
2	0
3	6
4	12
5	15
6	15
7	15
8	15
9	18
10	24
11	33
12	45
13	60
14	65
15	70
16	75
17	80
18	85
19	90
20	95
21	100
22	90
23	80
24	70
25	70

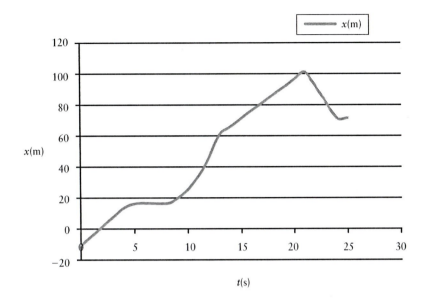

Figure 2-2 Problem 41

Between $t = 0$ s and $t = 4.5$ s:

$$x(t) = \frac{12 \text{ m} - (-12 \text{ m})}{4 \text{ s}}(t - (2 \text{ s})) = \left(6\frac{\text{m}}{\text{s}}\right)(t - (2 \text{ s}))$$

Between $t = 4.5$ s and $t = 8$ s: $x(t) = 15$ m.

Between $t = 8$ s and $t = 13$ s:

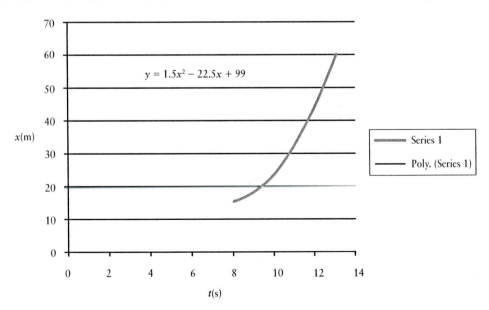

Figure 2-3 Problem 41

We can fit the data to a parabola in order to get position versus time in this region:

$$x(t) = \left(1.5\frac{m}{s^2}\right)t^2 - \left(22.5\frac{m}{s}\right)t + (99 \text{ m}) = \left(1.5\frac{m}{s^2}\right)(t - 7.5 \text{ s})^2 + (14.6 \text{ m})$$

Between $t = 13$ s and $t = 21$ s:

$$x(t) = \frac{100 \text{ m} - 60 \text{ m}}{8 \text{ s}}(t - 13 \text{ s}) + (60 \text{ m}) = \left(5\frac{m}{s}\right)(t - 13 \text{ s}) + (60 \text{ m})$$

Between $t = 21$ s and $t = 24$ s:

$$x(t) = \frac{70 \text{ m} - 100 \text{ m}}{3 \text{ s}}(t - 21 \text{ s}) + (100 \text{ m}) = \left(-10\frac{m}{s}\right)(t - 21 \text{ s}) + (100 \text{ m})$$

Between $t = 24$ s and $t = 25$ s: $x(t) = 70$ m.

Problems

2.47

SET UP

Kevin swims 4000 m in 1 hr. Because he ends at the same location as he starts, his total displacement is 0 m. This means his average velocity, which is his displacement divided by the time interval, is also zero. His average *speed*, on the other hand, is nonzero. The average speed takes the total distance covered into account.

SOLVE

Part a) Displacement = 0 m, so his $\boxed{\text{average velocity} = 0}$.

Part b)

$$v_{\text{average}} = \frac{4000 \text{ m}}{1 \text{ hr}} \times \frac{1 \text{ km}}{1000 \text{ m}} = \boxed{4\frac{km}{hr}}$$

Part c)

$$v_{\text{average}} = \frac{25 \text{ m}}{9.27 \text{ s}} \times \frac{1 \text{ km}}{1000 \text{ m}} \times \frac{3600 \text{ s}}{1 \text{ hr}} = \boxed{9.7\frac{km}{hr} = 2.7\frac{m}{s}}$$

REFLECT

Usually the terms "velocity" and "speed" are used interchangeably in everyday English, but the distinction between the two is important in physics. Kevin's average speed of 4 km/hr is about 2.5 mph. For comparison, Michael Phelps's record in the 100-m butterfly is 49.82 s, which gives him an average speed of 7.2 km/hr or 4.5 mph.

Get Help: Interactive Example – Interstate 74
P'Cast 2.2 – What's the Same

2.51

SET UP

It takes a sober driver 0.32 s to hit the brakes, while a drunk driver takes 1.0 s to hit the brakes. In both cases the car is initially traveling at 90 km/hr. Assuming it takes the same distance to come to a stop once the brakes are applied, the drunk driver travels for an extra $(1.0 - 0.32)$ s = 0.68 s at 90 km/hr before hitting the brakes. We can use the definition of average speed to calculate the extra distance the drunk driver travels.

SOLVE

$$\Delta x = (v_{average})(\Delta t) = \left(90\frac{km}{hr} \times \frac{1\ hr}{3600\ s} \times \frac{1000\ m}{1\ km}\right)(0.68\ s) = \boxed{17\ m}$$

REFLECT

The impaired driver travels an extra distance of over 55 ft before applying the brakes.

Get Help: Interactive Example – Interstate 74
P'Cast 2.2 – What's the Same

2.55

SET UP

Figure 2-4 Problem 55

We are given a plot of velocity versus time for a kangaroo rat. The displacement of the rat is related to the area under the velocity versus time curve. We can split the graph into simple, geometric shapes in order to easily approximate the area for the given time intervals. Area below the x-axis is considered to be negative.

SOLVE

Part a) 0 to 5 s

We will model this area as a rectangle:

$$\Delta x_{0-5\,s} = \left(0.25\frac{m}{s}\right)(5\ s) = \boxed{1.25\ m}$$

Part b) 0 to 10 s

We already know the area from 0–5 s. We just need to calculate the area from 5–10 s, which we can model as a trapezoid:

$$\Delta x_{0-10\,s} = \Delta x_{0-5\,s} + \Delta x_{5-10\,s} = (1.25\ m) + \frac{1}{2}((0.25\ m) + (0.2\ m))(5\ s)$$

$$= (1.25\ m) + (1.13\ m) = \boxed{2.4\ m}$$

Part c) 10 to 25 s

We will model this as a trapezoid:

$$\Delta x_{10-25 \, s} = \frac{1}{2}((0.2 \text{ m}) + (0.05 \text{ m}))(15 \text{ s}) = \boxed{1.9 \text{ m}}$$

Part d) 0 to 35 s

Parts (b) and (c) give us the area from 0–25 s. We can model the areas from 25–30 s and 30–35 s as triangles of equal area. The area from 30–35 s is negative.

$$\Delta x_{0-35 \, s} = \Delta x_{0-10 \, s} + \Delta x_{10-25 \, s} + \Delta x_{25-30 \, s} + \Delta x_{30-35 \, s}$$

$$= (2.4 \text{ m}) + (1.9 \text{ m}) + \frac{1}{2}\left(0.05\frac{\text{m}}{\text{s}}\right)(5 \text{ s}) - \frac{1}{2}\left(0.05\frac{\text{m}}{\text{s}}\right)(5 \text{ s}) = \boxed{4.3 \text{ m}}$$

REFLECT

Remember that areas below the x-axis are negative. In this case, the negative area corresponds to a negative displacement, which means the kangaroo rat is moving back toward its initial position. This makes sense because the velocity of the rat in this region is negative.

2.61

SET UP

A runner starts from rest and reaches a top speed of 8.97 m/s. Her acceleration is 9.77 m/s^2, which is a constant. We know her initial speed, her final speed, and her acceleration, and we are interested in the time it takes for her to reach that speed. We can rearrange $v_f = v_0 + at$ and solve for t.

SOLVE

$$v_f = v_0 + at$$

$$t = \frac{v_f - v_0}{a}$$

$$t = \frac{8.97\frac{\text{m}}{\text{s}} - 0}{9.77\frac{\text{m}}{\text{s}^2}} = \boxed{0.918 \text{ s}}$$

REFLECT

An acceleration of 9.77 m/s^2 means that her speed changes by 9.77 m/s every second. In this problem her speed changed a little less than that, so we expect the time elapsed to be a little less than a second.

2.65

SET UP

A Bugatti Veyron and Saleen S7 can accelerate from 0 to 60 mph in 2.4 s and 2.8 s, respectively. Assuming the acceleration of each car is constant, not only over that time interval but also for *any* time interval, we can approximate the acceleration of each car as the average acceleration. We first need to convert 60 mph into m/s in order to keep the units consistent. The acceleration is equal to the change in velocity divided by the change in time. We are interested in the distance each car travels when it accelerates from rest to 90 km/hr. We know the initial speed, final speed, and the acceleration, so we can use $v^2 - v_0^2 = 2a(\Delta x)$ to calculate Δx for each car. (See Problem 2.67 for a derivation of this equation.)

SOLVE
Finding the acceleration:

$$60\frac{\text{mi}}{\text{hr}} \times \frac{1.61 \text{ km}}{1 \text{ mi}} \times \frac{1 \text{ hr}}{3600 \text{ s}} \times \frac{1000 \text{ m}}{1 \text{ km}} = 27\frac{\text{m}}{\text{s}}$$

$$a_{\text{Bugatti}} = \frac{27\frac{\text{m}}{\text{s}} - 0}{2.4 \text{ s}} = 11.3\frac{\text{m}}{\text{s}^2}$$

$$a_{\text{Saleen}} = \frac{27\frac{\text{m}}{\text{s}} - 0}{2.8 \text{ s}} = 9.6\frac{\text{m}}{\text{s}^2}$$

Finding the distance:

$$v^2 - v_0^2 = 2a(\Delta x)$$

$$\Delta x = \frac{v^2 - v_0^2}{2a}$$

$$\Delta x_{\text{Bugatti}} = \frac{\left(90\frac{\text{km}}{\text{hr}} \times \frac{1000 \text{ m}}{1 \text{ km}} \times \frac{1 \text{ hr}}{3600 \text{ s}}\right)^2 - (0)^2}{2\left(11.3\frac{\text{m}}{\text{s}^2}\right)} = \boxed{28 \text{ m}}$$

$$\Delta x_{\text{Saleen}} = \frac{\left(90\frac{\text{km}}{\text{hr}} \times \frac{1000 \text{ m}}{1 \text{ km}} \times \frac{1 \text{ hr}}{3600 \text{ s}}\right)^2 - (0)^2}{2\left(9.6\frac{\text{m}}{\text{s}^2}\right)} = \boxed{33 \text{ m}}$$

REFLECT
These values are around 92 ft and 108 ft, respectively. A speed limit of 90 km/hr is around 55 mph. Most people don't drive expensive sports cars on the highway every day, so these distances are (obviously) much smaller than we should expect for most everyday cars.

Get Help: Interactive Example – Highway Patrol
P'Cast 2.7 – The Car, Again

2.69

SET UP
A sperm whale has an initial speed of 1 m/s and accelerates up to a final speed of 2.25 m/s at a constant rate of 0.1 m/s². Because we know the initial speed, final speed, and the acceleration, we can calculate the distance over which the whale travels by rearranging $v^2 - v_0^2 = 2a(\Delta x)$.

SOLVE

$$v^2 - v_0^2 = 2a(\Delta x)$$

$$\Delta x = \frac{v^2 - v_0^2}{2a} = \frac{\left(2.25\frac{m}{s}\right)^2 - \left(1\frac{m}{s}\right)^2}{2\left(0.1\frac{m}{s^2}\right)} = \boxed{20 \text{ m}}$$

REFLECT

This is a little longer than the average size of an adult male sperm whale (about 16 m). It will take the whale 12.5 s to speed up from 1 m/s to 2.25 m/s.

Get Help: Interactive Example – Highway Patrol

P'Cast 2.7 – The Car, Again

2.71

SET UP

The position of an object as a function of time is described by $x(t) = 12 - 6t + 3.2t^2$ in SI units. To calculate the displacement between $t = 4$ s and $t = 8$ s, we first need to find the position at each of those times and then calculate $\Delta x = x(t = 8 \text{ s}) - x(t = 4 \text{ s})$. The velocity as a function of time is the first derivative of the position with respect to time. Evaluating the derivative at $t = 3$ s will give the velocity at that time. After differentiating $x(t)$, we can set $v(t)$ equal to zero and solve for t. The acceleration is the first derivative of the velocity with respect to time or the second derivative of the position with respect to time.

SOLVE

Part a)

$$x(t = 8 \text{ s}) = 12 - 6(8) + 3.8(8)^2 = 169 \text{ m}$$

$$x(t = 4 \text{ s}) = 12 - 6(4) + 3.2(4)^2 = 39 \text{ m}$$

$$\Delta x = x(t = 8 \text{ s}) - x(t = 4 \text{ s}) = (169 \text{ m}) - (39 \text{ m}) = \boxed{130 \text{ m}}$$

Part b)

$$v(t) = \frac{d}{dt}x(t) = \frac{d}{dx}(12 - 6t + 3.2t^2) = \boxed{-6 + 6.4t} \text{ (SI units)}$$

$$v(t = 3 \text{ s}) = -6 + 6.4(3) = \boxed{13.2\frac{m}{s}}$$

Part c)

$$v(t) = -6 + 6.4t = 0$$

$$\boxed{t = 0.94 \text{ s}}$$

Part d)

$$a(t) = \frac{d}{dt}v(t) = \frac{d}{dx}(-6 + 6.4t) = \boxed{6.4\frac{m}{s^2}}$$

REFLECT

The object has a velocity of zero at $t = 0.94$ s, which means it changes direction at this point. Before this time, the object is moving toward negative x, stops, and then moves toward positive x. Its initial position, $x(0)$, is 12 m, while its position at $t = 0.94$ s is $x(0.94) = 9.2$ m and its position at $t = 2$ s is 12.8 m.

2.75

SET UP

Alex throws a ball straight down (toward $-y$) with an initial speed of $v_{0,G} = 4$ m/s from the top of a 50-m-tall tree. At the same instant, Gary throws a ball straight up (toward $+y$) with an initial speed of $v_{0,G}$ at a height of 1.5 m off the ground. (We will consider the ground to be $y = 0$, which makes the initial position of Alex's ball $y_{0,A} = 50$ m and the initial position of Gary's ball $y_{0,G} = 1.5$ m). Once Alex and Gary throw their respective balls, their accelerations will be equal to $a = -g$. We want to know how fast Gary has to throw his ball such that the two balls cross paths at $y_f = 25$ m at the same time.

We are given more information about Alex's throw than Gary's so that seems to be a reasonable place to start. First, we need to know the time at which Alex's ball is at $y_f = 25$ m. We can use $y_f = y_0 + v_0 t + \frac{1}{2}at^2$ with Alex's information to accomplish this. We know that, at this time, Gary's ball *also* has to be at $y_f = 25$ m. We can plug this time into $y_f = y_0 + v_0 t + \frac{1}{2}at^2$ again—this time with Gary's information—to calculate the initial speed of Gary's ball.

SOLVE

Time it takes Alex's ball to reach $y = 25$ m:

$$y_f = y_{0,A} + v_{0,A}t + \frac{1}{2}at^2 = y_{0,A} + v_{0,A}t + \frac{1}{2}(-g)t^2$$

$$-\frac{1}{2}gt^2 + v_{0,A}t + (y_{0,A} - y_f) = 0$$

This is a quadratic equation, so we will use the quadratic formula to find t:

$$t = \frac{-v_{0,A} \pm \sqrt{(v_{0,A})^2 - 4\left(-\frac{g}{2}\right)(y_{0,A} - y_f)}}{2\left(-\frac{g}{2}\right)} = \frac{-v_{0,A} \pm \sqrt{(v_{0,A})^2 + 4\left(\frac{g}{2}\right)(y_{0,A} - y_f)}}{-g}$$

$$= \frac{-\left(-4\frac{m}{s}\right) \pm \sqrt{\left(-4\frac{m}{s}\right)^2 + 4\left(\frac{9.8\frac{m}{s^2}}{2}\right)((50\ m) - (25\ m))}}{-\left(9.8\frac{m}{s^2}\right)} = \left(\frac{4 \pm 22.5}{-9.8}\right)\ s$$

The minus sign in the numerator gives the only physical answer:

$$t = \left(\frac{4 - 22.5}{-9.8}\right) s = 1.9 \ s$$

Speed Gary needs to launch the ball in order for it to be at $y = 25$ m at $t = 1.9$ s:

$$y_f = y_{0,G} + v_{0,G}t + \frac{1}{2}at^2 = y_{0,G} + v_{0,G}t + \frac{1}{2}(-g)t^2$$

$$v_{0,G} = \frac{y_f - y_{0,G} + \frac{1}{2}gt^2}{t} = \frac{(25 \ m) - (1.5 \ m) + \frac{1}{2}\left(9.8\frac{m}{s^2}\right)(1.9 \ s)^2}{1.9 \ s} = \boxed{21.7\frac{m}{s}}$$

REFLECT

Gary needs to throw his ball almost 50 mph in order for it to cross paths with Alex's at $y = 25$ m! The logic behind the way we solved this problem is much more important than the algebra used to solve this problem. Be sure every step makes logical sense; the crux of the argument is that the two balls have the same position at the same time.

Get Help: P'Cast 2.9 – How High?

2.77

SET UP

A person falls from a height of 6 ft off of the ground. We will call the ground $y = 0$. His initial speed is zero and acceleration is $-g$ because he is undergoing free fall. Using the conversion 1 ft = 0.3048 m and $v^2 = v_0^2 + 2a(\Delta y)$, we can calculate the speed at which he hits the ground.

SOLVE

Converting 6 ft into m:

$$6 \ ft \times \frac{0.3048 \ m}{1 \ ft} = 1.8 \ m$$

Solving for the final speed:

$$v^2 = v_0^2 + 2a(\Delta y) = v_0^2 + 2(-g)(\Delta y)$$

$$v = \sqrt{v_0^2 - 2g(\Delta y)} = \sqrt{0^2 - 2\left(9.8\frac{m}{s^2}\right)(0 - 1.8 \ m)} = \boxed{6\frac{m}{s}}$$

REFLECT

A speed of 6 m/s is about 13 mph. In addition to the speed consideration, it's much easier to topple from the top step than the lower ones, hence, the warning.

Get Help: P'Cast 2.11 – How Fast at Height?

2.81

SET UP

Figure 2-5 Problem 81

Mary is planning on dropping an apple out of her 17-m-high window to Bill. Bill is walking at a velocity of 2 m/s toward Mary's building and starts 120 m from directly below her window. Mary wants Bill to catch the apple, which means Bill and the apple need to be at the same location at the same time. We can use the constant acceleration equation for free fall to calculate the time it takes the apple to drop from an initial location of $y_0 = 17$ m to a final location of $y_f = 1.75$ m (presumably, Bill's height). Assume the apple has an initial velocity of zero. We can compare this to the time it takes Bill to walk 120 m in order to determine how long Mary should wait to drop the apple. Once we know how long Mary waits, we can find the distance Bill is from Mary since he is walking at a constant speed of 2 m/s. Bill's horizontal distance from Mary and the height Mary is in the air are two legs of a right triangle; the angle theta is related to these legs by the tangent.

SOLVE

Part a)

Time it takes Mary's apple to fall to a height of 1.75 m off the ground:

$$y_f = y_0 + v_0 t + \frac{1}{2}at^2 = y_0 + 0 + \frac{1}{2}(-g)t^2$$

$$t = \sqrt{\frac{2(y_f - y_0)}{-g}} = \sqrt{\frac{2((1.75 \text{ m}) - (17 \text{ m}))}{-\left(9.8\frac{\text{m}}{\text{s}^2}\right)}} = 1.76 \text{ s}$$

Time it takes Bill to walk 120 m:

$$(120 \text{ m})\left(\frac{1}{2\frac{\text{m}}{\text{s}}}\right) = 60 \text{ s}$$

Mary should wait $(60 \text{ s}) - (1.76 \text{ s}) = \boxed{58 \text{ s}}$ to drop her apple.

Part b) In 58 s, Bill travels $(58 \text{ s})\left(2\dfrac{\text{m}}{\text{s}}\right) = 116 \text{ m}$, which means he is $\boxed{4 \text{ m}}$ horizontally from the window.

Part c)

$$\tan(\theta) = \frac{4 \text{ m}}{17 \text{ m}}$$

$$\theta = \arctan\left(\frac{4}{17}\right) = \boxed{0.23 \text{ rad} = 13°}$$

17 m

4 m

Figure 2-6 Problem 81

REFLECT

As you would expect, Bill has to be reasonably close to Mary in order for him to catch the apple. If Mary throws the apple (that is, nonzero initial velocity), Bill would need to be even closer in order to catch the apple.

Get Help: P'Cast 2.9 – How High?
P'Cast 2.10 – How Long?

2.83

SET UP

Two trains are 300 m apart and traveling toward each other. Train 1 has an initial speed of 98 km/hr and an acceleration of -3.5 m/s^2. Train 2 has an initial speed of 120 km/hr and an acceleration of -4.2 m/s^2. To determine whether or not the trains collide, we can calculate the stopping distance required for each train, add them together, and see if it is less than 300 m. If so, the trains are safe; if not, the trains crash. We know the trains' initial speeds, final speeds, and acceleration, and we are interested in finding the distance, which suggests we use $v^2 = v_0^2 + 2a(\Delta x)$.

SOLVE

Stopping distance for train 1:

$$v^2 = v_0^2 + 2a(\Delta x)$$

$$\Delta x = \frac{v^2 - v_0^2}{2a} = \frac{0^2 - \left(98\dfrac{\text{km}}{\text{hr}} \times \dfrac{1 \text{ hr}}{3600 \text{ s}} \times \dfrac{1000 \text{ m}}{1 \text{ km}}\right)^2}{2\left(-3.5\dfrac{\text{m}}{\text{s}^2}\right)} = 105.8 \text{ m}$$

Stopping distance for train 2:

$$v^2 = v_0^2 + 2a(\Delta x)$$

$$\Delta x = \frac{v^2 - v_0^2}{2a} = \frac{0^2 - \left(120\dfrac{\text{km}}{\text{hr}} \times \dfrac{1 \text{ hr}}{3600 \text{ s}} \times \dfrac{1000 \text{ m}}{1 \text{ km}}\right)^2}{2\left(-4.2\dfrac{\text{m}}{\text{s}^2}\right)} = 132 \text{ m}$$

Total distance required to stop both trains: 105.8 m + 132 m = 237.8 m. This is less than the 300 m separating the trains, which means the trains will not collide. The distance separating the trains is 300 m − 237.8 m = $\boxed{62 \text{ m}}$.

REFLECT

As long as the trains are at least 237 m apart, they will not collide.

Get Help: Interactive Example – Highway Patrol

P'Cast 2.7 – The Car, Again

2.87

SET UP

The velocity of a rocket is given as a piecewise function. The SI units of velocity are m/s, so 125 and 116.1 have units of m/s. An exponent must be dimensionless, which means 0.12 has SI units of s^{-1}. The speed at any time is found by evaluating the function at that time; keep in mind that a piecewise function is defined differently over different time intervals. The displacement of the rocket is related to the integral of the velocity over the given time interval.

SOLVE

$$v(t) = \begin{cases} 125(1 - e^{-0.12t}) & 0 < t < 22 \text{ s} \\ 116.1 & t > 22 \text{ s} \end{cases}$$

Part a) The numbers 125 and 116.1 both have SI units of $\boxed{\text{m/s}}$ because those are the SI units for velocity. The exponent of e needs to be dimensionless; t has dimensions of time, which means 0.12 has dimensions of reciprocal time. The SI unit associated with 0.12 is, therefore, $\boxed{s^{-1}}$.

Part b)

Speed at $t = 2.2$ s:

$$v(t = 2.2 \text{ s}) = 125(1 - e^{-0.12(2.2)}) = \boxed{29\frac{\text{m}}{\text{s}}}$$

Speed at $t = 25$ s:

$$v(t = 25 \text{ s}) = \boxed{116.1\frac{\text{m}}{\text{s}}}$$

Part c)

$$\Delta x = \int_0^{10 \text{ s}} v(t)dt = \int_0^{10 \text{ s}} (125(1 - e^{-0.12t}))dt = \left[125\left(t + \frac{1}{0.12}e^{-0.12t}\right)\right]_0^{10 \text{ s}}$$

$$= 125\left[10 + \frac{1}{0.12}e^{-0.12(10)} - 0 - \frac{1}{0.12}e^{-0.12(0)}\right] = 125\left[10 + \frac{1}{0.12}e^{-1.2} - \frac{1}{0.12}\right] = \boxed{520 \text{ m}}$$

Part d)

$$\Delta x = \int_0^{30 \text{ s}} v(t)dt = \int_0^{22 \text{ s}} (125(1 - e^{-0.12t}))dt + \int_{22 \text{ s}}^{30 \text{ s}} 116.1 dt = \left[125\left(t + \frac{1}{0.12}e^{-0.12t}\right)\right]_0^{22 \text{ s}} + [116.1t]_{22 \text{ s}}^{30 \text{ s}}$$

$$= 125\left[22 + \frac{1}{0.12}e^{-0.12(22)} - 0 - \frac{1}{0.12}e^{-0.12(0)}\right] + [116.1(30) - 116.1(22)]$$

$$= 125\left[22 + \frac{1}{0.12}e^{-2.64} - \frac{1}{0.12}\right] + [928.8] = \boxed{2700 \text{ m}}$$

REFLECT

Be aware of the time intervals when working with piecewise functions. The velocity function for the rocket changes at $t = 22$ s, which is why we had to split the integral in part (d) into two parts.

2.93

SET UP

A ball is dropped from an unknown height above your window. You observe that the ball takes 0.18 s to traverse the length of your window, which is 1.5 m. We can calculate the velocity of the ball when it is at the top of the window from the length of the window, the acceleration due to gravity, and the time it takes to pass by the window. Once we have the velocity at that point, we can determine the height the ball needed to fall to achieve that speed, assuming that its initial velocity was zero. Throughout the problem, we will define *down* to be negative.

SOLVE

Speed of ball at top of window:

$$\Delta y = v_0 t + \frac{1}{2}at^2 = v_0 t + \frac{1}{2}(-g)t^2$$

$$v_0 = \frac{\Delta y + \frac{1}{2}gt^2}{t} = \frac{(-1.5 \text{ m}) + \frac{1}{2}\left(9.8\frac{\text{m}}{\text{s}^2}\right)(0.18 \text{ s})^2}{(0.18 \text{ s})} = -7.45\frac{\text{m}}{\text{s}}$$

Distance the ball drops to achieve a speed of 7.45 m/s:

$$v^2 - v_0^2 = 2a(\Delta y) = 2(-g)(\Delta y)$$

$$\Delta y = \frac{v^2 - v_0^2}{2(-g)} = \frac{\left(-7.45\frac{\text{m}}{\text{s}}\right)^2 - 0}{2\left(-9.8\frac{\text{m}}{\text{s}^2}\right)} = -2.8 \text{ m}$$

The ball started at a distance of $\boxed{2.8 \text{ m}}$ above your window.

REFLECT

Be careful with the signs of Δy, a, and v in this problem.

Get Help: P'Cast 2.9 – How High?

2.97

SET UP

A rocket with two stages of rocket fuel is launched straight up into the air from rest. Stage 1 lasts 10.0 s and provides a net upward acceleration of 15 m/s^2. Stage 2 lasts 5.0 s and provides a net upward acceleration of 12 m/s^2. After Stage 2 finishes, the rocket continues to travel upward under the influence of gravity alone until it reaches its maximum height and then falls back toward Earth. We can split the rocket's flight into four parts: (1) Stage 1, (2) Stage 2, (3) between the end of Stage 2 and reaching the maximum height, and (4) falling from the maximum height back to Earth's surface. The acceleration of the rocket is constant over each of these legs, so we can use the constant acceleration equations to determine the total distance covered in each leg and the duration of each leg. The initial speed for legs #1 and #4 is zero, but we will need to calculate the initial speeds for legs #2 and #3. The maximum altitude is equal to the distance covered in legs #1–#3; the time required for the rocket to return to Earth is equal to the total duration of its flight, which is the sum of the durations of legs #1–#4.

SOLVE

Maximum altitude
Distance traveled in Stage 1:

$$\Delta y = v_0 t + \frac{1}{2}at^2$$

$$\Delta y = 0 + \frac{1}{2}\left(15\frac{m}{s^2}\right)(10.0 \text{ s})^2 = 750 \text{ m}$$

Speed after Stage 1:

$$v = v_0 + at = 0 + \left(15\frac{m}{s^2}\right)(10.0 \text{ s}) = 150\frac{m}{s}$$

Distance traveled in Stage 2:

$$\Delta y = v_0 t + \frac{1}{2}at^2$$

$$\Delta y = \left(150\frac{m}{s}\right)(5.0 \text{ s}) + \frac{1}{2}\left(12\frac{m}{s^2}\right)(5.0 \text{ s})^2 = 900 \text{ m}$$

Speed after Stage 2:

$$v = v_0 + at = \left(150\frac{m}{s}\right) + \left(12\frac{m}{s^2}\right)(5.0 \text{ s}) = 210\frac{m}{s}$$

Distance traveled after Stage 2:

$$v^2 - v_0^2 = 2a(\Delta y)$$

$$\Delta y = \frac{v^2 - v_0^2}{2a} = \frac{0 - \left(210\frac{m}{s}\right)^2}{2\left(-9.8\frac{m}{s^2}\right)} = 2250 \text{ m}$$

Maximum altitude: 750 m + 900 m + 2250 m = $\boxed{3900 \text{ m}}$.

Time required to return to the surface
Time after Stage 2:

$$v = v_0 + at$$

$$t = \frac{v - v_0}{a} = \frac{0 - \left(210\frac{m}{s}\right)}{\left(-9.8\frac{m}{s^2}\right)} = 21.4 \text{ s}$$

Time from maximum height to the ground:

$$\Delta y = v_0 t + \frac{1}{2}at^2 = 0 + \frac{1}{2}(-g)t^2$$

$$t = \sqrt{\frac{2(\Delta y)}{-g}} = \sqrt{\frac{2(-3900 \text{ m})}{-\left(9.8\frac{\text{m}}{\text{s}^2}\right)}} = 28.2 \text{ s}$$

Total time of flight: $10.0 \text{ s} + 5.0 \text{ s} + 21.4 \text{ s} + 28.2 \text{ s} = \boxed{64.6 \text{ s}}$.

REFLECT

An altitude of 3900 m is around 2.5 mi. The simplest way of solving this problem was to split it up into smaller, more manageable calculations and then put all of the information together.

Get Help: P'Cast 2.9 – How High?

P'Cast 2.10 – How Long?

Chapter 3
Motion in Two Dimensions

Conceptual Questions

3.1 Part a) No, the sum of two vectors that have different magnitudes can never equal zero. The only time two vectors can add together to have zero magnitude is when they point in opposite directions and have the same magnitude.

Part b) Yes, the sum of three (or more) vectors with different magnitudes can be equal to zero.

Get Help: Picture It – Adding and Subtracting Vectors

3.5 Air resistance can be modeled as a force that depends on the velocity. The greater the speed, the greater the magnitude of the drag force. Also, the drag force points in the direction opposite to the direction of motion. Therefore, when the effects of air resistance are taken into account, the projectile's speed in the vertical direction will decrease more rapidly as it ascends and increase less rapidly when it comes down compared to when the effects of air resistance are ignored. When effects of air resistance are ignored, a projectile experiences no acceleration in the horizontal direction and, therefore, travels at constant speed horizontally. Due to the drag force resulting from air resistance, however, a projectile experiences a horizontal acceleration that causes its horizontal speed to decrease.

3.9 Part a) Maximum range is achieved at a launch angle of 45 degrees.

Part b) The longest time of flight is achieved by launching the projectile at an angle of 90 degrees (that is, straight up). We can see this from Equation 3-23: $t_{peak} = \dfrac{v_0 \sin(\theta)}{g}$.

Part c) The greatest height is achieved by launching the projectile at an angle of 90 degrees (that is, straight up). We can see this from Equation 3-24:

$$(y - y_0)_{peak} = \frac{1}{2}\frac{v_0^2 \sin^2(\theta)}{g}.$$

Get Help: P'Cast 3.4 – How High?
P'Cast 3.5 – How Far?
P'Cast 3.6 – How Long?

3.13 Yes, the ape is accelerating at the bottom of its swing. The force from the vine is accelerating the ape upward since it is at the bottom of the swing that it changes from traveling downward to traveling upward.

Multiple-Choice Questions

3.17 B (45 degrees). $\theta = \arctan\left(\dfrac{A_y}{A_x}\right) = \arctan(1) = 45°$.

3.21 D ($d_2 = 4d_1$).

$$d_1 = \frac{2v_0^2 \sin(\theta)\cos(\theta)}{g}$$

$$d_2 = \frac{2(2v_0)^2 \sin(\theta)\cos(\theta)}{g} = 4\frac{2v_0^2 \sin(\theta)\cos(\theta)}{g} = 4d_1$$

Get Help: Interactive Example – Baseball
P'Cast 3.5 – How Far?

Estimation Questions

3.29

Figure 3-1 Problem 29

The ball reached a maximum height of $\Delta y = 110$ ft:

$$v_y^2 - v_{0,y}^2 = 2a_y(\Delta y)$$

$$v_{0,y} = \sqrt{-2a_y(\Delta y)} = \sqrt{-2\left(-32\frac{\text{ft}}{\text{s}^2}\right)(110 \text{ ft})} = 26.5\frac{\text{ft}}{\text{s}}$$

Estimating the initial launch angle of the ball from the graph to be 30 degrees:

$$v_{0,y} = v_0\sin(30°) = \frac{v_0}{2}$$

$$v_0 = 2v_{0,y} = 2\left(26.5\frac{\text{ft}}{\text{s}}\right) = 53\frac{\text{ft}}{\text{s}}$$

But the air resistance is not completely negligible here. How much is there? The peak occurs at a horizontal distance of 300 ft and the ball hits Earth at around 540 ft. Without air resistance, the ball should have landed at around 600 ft, so 20% of the forward progress is lost. Therefore, a reasonable estimate of the initial speed is (1.2) (53 ft/s) = 64 ft/s.

Get Help: Interactive Example – Baseball

Problems

3.33

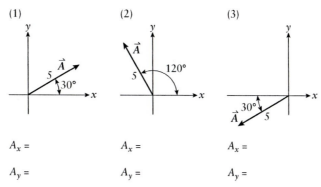

Figure 3-2 Problem 33

SET UP

A vector has a magnitude of 5 and is shown in three different orientations. Since we are given the angles in each case, we can calculate the components directly using $A_x = A\cos(\theta)$ and $A_y = A\sin(\theta)$.

SOLVE

(1)
$$A_x = 5\cos(30°) = \boxed{4.3}$$
$$A_y = 5\sin(30°) = \boxed{2.5}$$

(2)
$$A_x = 5\cos(120°) = -5\cos(60°) = \boxed{-2.5}$$
$$A_y = 5\sin(120°) = 5\sin(60°) = \boxed{4.3}$$

(3)
$$A_x = 5\cos(210°) = -5\cos(30°) = \boxed{-4.3}$$
$$A_y = 5\sin(210°) = -5\sin(30°) = \boxed{2.5}$$

REFLECT

It is a good idea to redraw the right triangle each time to ensure you are taking the sine or cosine of the correct angle.

3.37

SET UP

Two vectors are given. We can calculate their difference by subtracting their components and then determining the magnitude and angle using the Pythagorean theorem and the tangent, respectively.

SOLVE

$$\vec{A} = (2.00)\hat{x} + (6.00)\hat{y}$$

$$\vec{B} = (3.00)\hat{x} - (2.00)\hat{y}$$

$$\vec{D} = \vec{A} - \vec{B} = ((2.00)\hat{x} + (6.00)\hat{y}) - ((3.00)\hat{x} - (2.00)\hat{y}) = (-1.00)\hat{x} + (8.00)\hat{y}$$

$$D = \sqrt{(-1.00)^2 + (8.00)^2} = \boxed{8.06}$$

$$\theta = \arctan\left(\frac{8.00}{-1.00}\right) = \boxed{1.70}$$

REFLECT

An angle of 1.70 radians is equal to 97.1 degrees. The difference $\vec{A} - \vec{B}$ is the same as the sum $\vec{A} + (-1)\vec{B}$.

3.41

SET UP

The initial and final velocity vectors of an object are given. We can calculate the average acceleration vector of the object by calculating the change in velocity, dividing this new vector by 5 s, and then finding its magnitude. We are not given the intermediate velocity vectors, so we cannot determine if the acceleration was uniform.

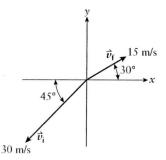

Figure 3-3 Problem 41

SOLVE

$$v_{i, x} = -\left(30\frac{m}{s}\right)\cos(45°) = -21\frac{m}{s}$$

$$v_{i, y} = -\left(30\frac{m}{s}\right)\sin(45°) = -21\frac{m}{s}$$

$$\vec{v}_i = -\left(21\frac{m}{s}\right)\hat{x} - \left(21\frac{m}{s}\right)\hat{y}$$

$$v_{f, x} = \left(15\frac{m}{s}\right)\cos(30°) = 13\frac{m}{s}$$

$$v_{f, y} = \left(15\frac{m}{s}\right)\sin(30°) = 7.5\frac{m}{s}$$

$$\vec{v}_f = \left(13\frac{m}{s}\right)\hat{x} + \left(7.5\frac{m}{s}\right)\hat{y}$$

$$\Delta\vec{v} = \vec{v}_f - \vec{v}_i = \left(\left(13\frac{m}{s}\right)\hat{x} + \left(7.5\frac{m}{s}\right)\hat{y}\right) - \left(-\left(21\frac{m}{s}\right)\hat{x} - \left(21\frac{m}{s}\right)\hat{y}\right)$$

$$= \left(34\frac{m}{s}\right)\hat{x} + \left(28.5\frac{m}{s}\right)\hat{y}$$

$$\vec{a}_{average} = \frac{\Delta\vec{v}}{\Delta t} = \frac{\left(34\frac{m}{s}\right)\hat{x} + \left(28.5\frac{m}{s}\right)\hat{y}}{5\ s} = \left(6.8\frac{m}{s^2}\right)\hat{x} + \left(5.7\frac{m}{s^2}\right)\hat{y}$$

$$a_{average} = \sqrt{\left(6.8\frac{m}{s^2}\right)^2 + \left(5.7\frac{m}{s^2}\right)^2} = \boxed{8.9\frac{m}{s^2}}$$

No, it's not possible to know whether the acceleration was uniform from the information given. We would need to know the acceleration as a function of time in order to answer that question.

REFLECT

The average acceleration vector points in the same direction as the change in the velocity. Its magnitude is scaled by the time interval.

3.47

SET UP

Five balls are thrown off the edge of a cliff with the same speed but at different angles. We can use intuition and the constant acceleration equations to determine which ball travels the farthest horizontal distance, which ball takes the longest to hit the ground, and which ball has the greatest speed when landing.

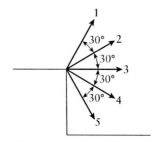

Figure 3-4 Problem 47

SOLVE

Part a) Balls 1 and 2 should travel the farthest since they are thrown upward. Balls 4 and 5 should travel the shortest distance since they are thrown downward. The initial x velocity for ball 2 is larger than ball 1, so it should travel farther horizontally before hitting the ground. Therefore, ball 2 travels the farthest followed by balls 1, 3, 4, and 5.

Part b) Ball 1 should take the longest to hit the ground because it has the largest, positive initial y velocity, followed by balls 2, 3, 4, and 5.

Part c)

$$v^2 - v_0^2 = 2a_y(\Delta y)$$

$$v = \sqrt{v_0^2 + 2a_y(\Delta y)}$$

The initial speed, the acceleration, and the height traveled are all the same for the five cases. Therefore, all five balls will land with the same speed.

REFLECT

Be careful when searching through the text for equations to use. Some of them will only work for specific cases and not in general. For example, Equation 3-26 $\left((x - x_0)_{\text{range}} = \frac{2v_0^2 \sin(\theta)\cos(\theta)}{g} \right)$ can only be used for a projectile returning to its initial launch height.

3.49

SET UP

A tiger leaps out of a tree that is 4.00 m tall. He lands 5.00 m from the base of the tree. The tiger's initial velocity is completely in the horizontal direction, which means the y component of his initial velocity is zero. We will choose the tiger's starting point to be the origin and up to be positive y. We can assume the tiger is jumping in the positive x direction. We can use the *y* component information to solve for *t* in terms of known quantities and then solve for the initial speed.

SOLVE

Know/Don't Know table:

	X		Y
x_0	0 m	y_0	0 m
x_f	5.000 m	y_f	−4.00 m
v_{0x}	WANT	v_{0y}	0
v_x	?	v_y	?
a_x	0	a_y	−9.8 m/s^2
t	?	t	?

Solving for initial speed:

$$\Delta y = v_{0,y}t + \frac{1}{2}a_y t^2 = 0 + \frac{1}{2}a_y t^2$$

$$t = \sqrt{\frac{2(\Delta y)}{a_y}}$$

$$\Delta x = v_{0,x}t + \frac{1}{2}a_x t^2 = v_{0,x}t + 0 = v_{0,x}\left(\sqrt{\frac{2(\Delta y)}{a_y}}\right)$$

$$v_{0,x} = \Delta x\left(\sqrt{\frac{a_y}{2(\Delta y)}}\right) = (5.00 \text{ m})\left(\sqrt{\frac{-9.81\frac{m}{s^2}}{2(-4.00 \text{ m})}}\right) = \boxed{5.53\frac{m}{s}}$$

REFLECT

We could have chosen the base of the tree to be the origin rather than the tiger's initial location. This would not have affected the sign or magnitude of the *y* displacement.

3.53

SET UP

A ball is attached to a 0.870-m-long string and is moving in a circle with a constant speed of 3.36 m/s. The ball is undergoing uniform circular motion, so its acceleration has a magnitude of $a = \frac{v^2}{R}$ and points toward the center of the circle.

SOLVE

$$a = \frac{v^2}{R} = \frac{\left(3.36\frac{m}{s}\right)^2}{0.870 \text{ m}} = \boxed{13.0\frac{m}{s^2} \text{ inward}}$$

REFLECT

Centripetal acceleration always points toward the center of the circle.

3.57

SET UP

A Ferris wheel is 76 m in diameter and completes a revolution in 20 min, which means a point on the rim will travel a length of one circumference in 20 min. We will assume the Ferris wheel is undergoing uniform circular motion, so its acceleration is equal to $a = \dfrac{v^2}{R}$.

SOLVE

$$v = \frac{1 \text{ rev}}{20 \text{ min}} \times \frac{1 \text{ min}}{60 \text{ s}} \times \frac{\pi(76 \text{ m})}{1 \text{ rev}} = 0.199 \frac{\text{m}}{\text{s}}$$

$$a = \frac{v^2}{R} = \frac{\left(0.199 \dfrac{\text{m}}{\text{s}}\right)^2}{\left(\dfrac{76 \text{ m}}{2}\right)} = \boxed{1.0 \times 10^{-3} \frac{\text{m}}{\text{s}^2}}$$

REFLECT

We are given the *diameter* of the wheel, not the *radius*. The radial acceleration uses the radius, so be sure to divide by 2 when performing the calculation.

3.61

SET UP

The space shuttle is in an orbit about 300 km above the surface of Earth, which has a radius of 6.38×10^6 m. We will assume the orbit is circular and that the space shuttle is moving with a constant speed. The shuttle sweeps out a circle of radius $R = R_{\text{Earth}} + R_{\text{orbit}}$ and covers a distance of $2\pi R$ in one period T. The acceleration of the shuttle is $a = \dfrac{v^2}{R}$ and points toward the center of Earth.

SOLVE

$$v = \frac{2\pi(R_{\text{Earth}} + R_{\text{orbit}})}{T} = \frac{2\pi((6.38 \times 10^6 \text{ m}) + (3.00 \times 10^5 \text{ m}))}{(5.43 \times 10^3 \text{ s})} = 7730 \frac{\text{m}}{\text{s}}$$

$$a = \frac{v^2}{R_{\text{Earth}} + R_{\text{orbit}}} = \frac{\left(7730 \dfrac{\text{m}}{\text{s}}\right)^2}{(6.38 \times 10^6 \text{ m}) + (3.00 \times 10^5 \text{ m})}$$

$$= \boxed{8.9 \frac{\text{m}}{\text{s}^2} \text{ toward the center of Earth}}$$

REFLECT

This is about 90% of the acceleration due to gravity at Earth's surface.

Get Help: P'Cast 3.8 – Artificial Gravity

3.65

SET UP

A peregrine falcon pulls out of a dive and into a circular arc at a speed of 100 m/s. It experiences an acceleration of 0.6g, where $g = 9.8$ m/s^2. Assuming the bird's speed is constant, we can calculate the radius of the turn from the centripetal acceleration.

SOLVE

$$a = \frac{v^2}{R}$$

$$R = \frac{v^2}{a} = \frac{\left(100\frac{m}{s}\right)^2}{(0.6)\left(9.8\frac{m}{s^2}\right)} = 1700 \text{ m} = \boxed{1.7 \text{ km}}$$

REFLECT

Peregrine falcons are known to be one of the fastest animals on Earth. A speed of 100 m/s is roughly 220 mph!

3.69

SET UP

Steve wants to throw a football to Jerry at an initial speed of 15.0 m/s at an angle of 45 degrees. Jerry runs past Steve at a speed of 8.00 m/s in the direction the football will be thrown. In order for Jerry to catch the football, Jerry and the football need to be at the same place at the same time. We can calculate the distance the football will travel and the time it takes for it to travel that distance. Because we are told that the football is caught at the same height from which it is released, we can use the maximum horizontal range equation (Equation 3-26). We can compare this time with the time it takes Jerry to run the same distance. The difference between these times is how long Steve should wait to throw the ball.

In part (b), Steve starts to run with the football at a speed of 1.50 m/s down the field as well. This means we need to use the speeds for the football and for Jerry relative to Steve. The football will still cover 23.0 m since its speed *relative to the ground* is still 15.0 m/s.

SOLVE

Part a)

Distance the football travels:

$$(x - x_0)_{max} = \frac{2v_0^2 \sin(\theta)\cos(\theta)}{g} = \frac{2\left(15.0\frac{m}{s}\right)^2 \sin(45°)\cos(45°)}{9.81\frac{m}{s^2}} = 23.0 \text{ m}$$

Time it takes the football to travel 23.0 m:

$$\Delta x = v_{0,x}t = v_0\cos(45°)t$$

$$t = \frac{\Delta x}{v_0\cos(45°)} = \frac{23.0 \text{ m}}{\left(15.0\frac{\text{m}}{\text{s}}\right)\cos(45°)} = 2.17 \text{ s}$$

Time it takes Jerry to run 23.0 m:

$$v_J = \frac{\Delta x}{\Delta t}$$

$$\Delta t = \frac{\Delta x}{v_J} = \frac{23.0 \text{ m}}{8.00\frac{\text{m}}{\text{s}}} = 2.88 \text{ s}$$

Steve should wait $2.88 \text{ s} - 2.17 \text{ s} = \boxed{0.71 \text{ s}}$ to throw the football.

Part b)

Time it takes the football to travel 23.0 m at a relative speed of 13.5 m/s:

$$\Delta x = v_{0,x}t = v_0\cos(45°)t$$

$$t = \frac{\Delta x}{v_0\cos(45°)} = \frac{23.0 \text{ m}}{\left(13.5\frac{\text{m}}{\text{s}}\right)\cos(45°)} = 2.41 \text{ s}$$

Time it takes Jerry to run 23.0 m at a relative speed of 6.5 m/s:

$$v_J = \frac{\Delta x}{\Delta t}$$

$$\Delta t = \frac{\Delta x}{v_J} = \frac{23.0 \text{ m}}{6.50\frac{\text{m}}{\text{s}}} = 3.53 \text{ s}$$

Steve should wait $3.53 \text{ s} - 2.41 \text{ s} = \boxed{1.12 \text{ s}}$ to throw the football.

REFLECT

It makes sense that Steve should wait longer to release the football when he is running rather than standing still.

Get Help: P'Cast 3.6 – How Long?

3.75

SET UP

A water balloon is thrown horizontally at an initial speed of 2.00 m/s from the top of a 6.00-m-tall building. At the same time, a second water balloon is thrown straight down from the same height at an initial speed of 2.00 m/s. To determine which balloon hits the ground first, we need to calculate the time it takes each balloon to hit the ground; whichever balloon has the shorter time will hit the ground first. Since the balloons start with the same initial speed and fall the same distance, they will hit the ground with the same final speed.

SOLVE

Know/Don't Know table for balloon 1:

	X		Y
x_0	0 m	y_0	6.00 m
x_f	?	y_f	0
v_{0x}	2.00 m/s	v_{0y}	0
v_x	?	v_y	?
a_x	0	a_y	−9.81 m/s^2
t	?	t	?

Time for balloon 1 to hit the ground:

$$\Delta y = v_{0,y}t + \frac{1}{2}a_y t^2 = 0 + \frac{1}{2}a_y t^2$$

$$t = \sqrt{\frac{2(\Delta y)}{a_y}} = \sqrt{\frac{2(-6.00 \text{ m})}{\left(-9.81\frac{\text{m}}{\text{s}^2}\right)}} = 1.106 \text{ s}$$

Know/Don't Know table for balloon 2:

	Y
y_0	6.00 m
y_f	0
v_{0y}	−2.00 m/s
v_y	?
a_y	−9.81 m/s^2
t	?

Time for balloon 2 to hit the ground:

$$\frac{1}{2}a_y t^2 + v_{0,y}t - \Delta y = 0$$

$$t = \frac{-v_{0,y} \pm \sqrt{v_{0,y}^2 - 4\left(\frac{1}{2}a_y\right)(-\Delta y)}}{2\left(\frac{1}{2}a_y\right)} = \frac{-\left(-2.00\frac{\text{m}}{\text{s}}\right) \pm \sqrt{\left(-2.00\frac{\text{m}}{\text{s}}\right)^2 + 2\left(-9.81\frac{\text{m}}{\text{s}^2}\right)(-6.00 \text{ m})}}{\left(-9.81\frac{\text{m}}{\text{s}^2}\right)}$$

$$= \frac{-2.00 \pm 11.0}{-9.81} \text{ s}$$

Taking the negative root: $t = 0.917$ s.

Balloon 2 lands 0.189 s before balloon 1 .

Final speed of the balloons:

$$v^2 - v_0^2 = 2a_y(\Delta y)$$

$$v = \sqrt{v_0^2 + 2a_y(\Delta y)}$$

Since the balloons start with the same initial speed and fall the same distance, the balloons will $\boxed{\text{land with the same speed}}$.

REFLECT

It will save you time and frustration in the long run if you solve each problem algebraically first and then plug in numbers at the end. For example, we could easily see the balloons have the same final speed from the general, algebraic answer in one calculation, rather than two.

Get Help: P'Cast 3.6 – How Long?

3.79

SET UP

Gabriele Reinsch threw a discus 76.80 m. She launched it an angle of 45 degrees from a height of 2.0 m above the ground. We will take Gabriele's feet to be the origin, the positive x direction to be the direction she threw the discus, and up to be the positive y direction. We can use the x information to solve for t and plug this into the y equation to solve for her initial speed. Remember that $\cos(45°) = \sin(45°) = \dfrac{1}{\sqrt{2}}$; this will help simplify the algebra.

SOLVE

Know/Don't Know table:

	X		Y
x_0	0 m	y_0	2.0 m
x_f	76.80 m	y_f	0 m
v_{0x}	$v_0\cos(45°)$	v_{0y}	$v_0\sin(45°)$
v_x	?	v_y	?
a_x	0	a_y	-9.81 m/s^2
t	?	t	?

$$\Delta x = v_{0,x}t + \frac{1}{2}a_x t^2 = v_0\cos(45°)t + 0$$

$$t = \frac{\Delta x}{v_0\cos(45°)}$$

$$\Delta y = v_{0,y}t + \frac{1}{2}a_y t^2 = v_0\sin(45°)\left(\frac{\Delta x}{v_0\cos(45°)}\right) + \frac{1}{2}a_y\left(\frac{\Delta x}{v_0\cos(45°)}\right)^2 = \Delta x + a_y\left(\frac{\Delta x}{v_0}\right)^2$$

$$v_0 = \sqrt{\frac{-a_y(\Delta x)^2}{(\Delta x) - (\Delta y)}} = \sqrt{\frac{-\left(-9.81\dfrac{m}{s^2}\right)(76.80\ m)^2}{(76.80\ m)-(-2.0\ m)}} = \boxed{27.1\dfrac{m}{s}}$$

REFLECT

We cannot use the maximum horizontal range formula from the text because the discus does not land at the same height it was released.

3.85

SET UP

A sample in a centrifuge initially experiences a radial acceleration of $a_1 = 100g$. The acceleration then increases by a factor of 8, $a_2 = 800g$. We can determine the new rotation speed in terms of the original rotation speed by looking at the ratio of the two radial accelerations. The radius is the same in both cases, since the size of the centrifuge doesn't change.

SOLVE

$$a_1 = 100g = \frac{v_1^2}{R}$$

$$a_2 = 800g = \frac{v_2^2}{R}$$

$$\frac{a_2}{a_1} = \frac{800g}{100g} = 8 = \frac{\left(\dfrac{v_1^2}{R}\right)}{\left(\dfrac{v_2^2}{R}\right)} = \frac{v_1^2}{v_2^2}$$

$$\boxed{v_2 = v_1\sqrt{8} \approx 2.83v_1}$$

REFLECT

If a problem asks you to find the factor by which a quantity increases or decreases, first try taking a ratio of the quantities. Usually some terms will remain constant over the process and will cancel out in your ratio.

Chapter 4
Newton's Laws of Motion

Conceptual Questions

4.1 Yes, the direction of the net force will determine the direction of the acceleration vector. Mass is a scalar quantity and will not affect the direction.

4.7 You can double the rope up. Each of the two strands pulls only 425 N.

4.9 Yes, both the Sun and Earth pull each other and make up a force pair associated with the gravitational interaction.

4.15 The net force on the bathroom scale is zero since it is at rest.

4.19 It's easier to lift a truck on the Moon. Weight is due to gravity. The Moon has less gravity than Earth.

4.23 A bird begins flying by pushing with its wings backward against the air. The air then exerts an equal and opposite force forward on the bird, and it is this force on the bird that propels it forward.

Multiple-Choice Questions

4.25 B (rear). The car accelerates forward quickly due to the rear-end collision, while the person stays in place.

4.31 D (at rest or in motion with a constant velocity). A net force of zero means the acceleration is also zero.

Estimation Questions

4.37 A baseball throw can be in the range of 40 m/s (about 89 mph). A baseball has a mass of around 0.15 kg. The powerful part of the throw lasts about 0.05 s. This gives

$$F = m\frac{\Delta v}{\Delta t} = (0.15 \text{ kg})\frac{\left(40\frac{\text{m}}{\text{s}}\right)}{0.05 \text{ s}} \approx 120 \text{ N}.$$

4.45 The weight of the rider will reach its maximum value during the ascent from the ground floor to the fourth floor. This is because the weight of the rider will be added to the force required to accelerate from 0 m/s to 10 m/s.

Figure 4-1 Problem 45

Get Help: Picture it – Apparent Weight

Problems

4.49

SET UP

The engine of a 1250-kg car can deliver a maximum force of 15,000 N. Since this is the only force acting on the car, we can set this force equal to the mass times the acceleration and solve for the magnitude of the acceleration.

SOLVE

$$\sum F = ma$$

$$a = \frac{\sum F}{m} = \frac{15,000 \text{ N}}{1250 \text{ kg}} = \boxed{12\frac{\text{m}}{\text{s}^2}}$$

REFLECT

Because there is only one force acting on the car, this force is equal to the net force.

4.53

SET UP

A tuna has a mass of 250 kg. Its weight is equal to the magnitude of the force of gravity on it, which is equal to the fish's mass multiplied by the acceleration due to gravity.

SOLVE

$$W = mg = (250 \text{ kg})\left(9.8\frac{\text{m}}{\text{s}^2}\right) = \boxed{2450 \text{ N}}$$

REFLECT

This is a weight of 550 lb.

4.57

SET UP

A person is pushing a box across a smooth floor at a steadily increasing speed, which means the box is accelerating along that direction. The force due to the man on the box acts horizontally, while the normal force and the force due to gravity act vertically.

SOLVE

Figure 4-2 Problem 57

REFLECT

The normal force and force due to gravity are equal in magnitude because the box is not accelerating in the vertical direction.

4.61

SET UP

Three rugby players are pulling horizontally on ropes attached to a stationary box. Player 1 pulls with a force of 100 N at an angle of −60 degrees. Player 2 pulls with a force of 200 N at an angle of +37 degrees. We can use Newton's second law to calculate the components of Player 3's force. Since the box is at rest, the acceleration is equal to zero. Once we know the components of Player 3's force, we can calculate its magnitude and direction and draw it on the figure.

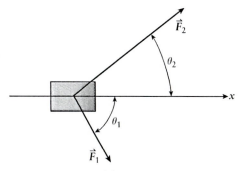

Figure 4-3 Problem 61

Player 3's rope breaks, which means $F_3 = 0$ N, and Player 2 compensates for this by pulling with a force of only 150 N. The direction of the box's resulting acceleration is equal to the direction of the net force due to Players 1 and 2. We can calculate the mass of the box from Newton's second law.

SOLVE

Part a)

$$\sum F_x = F_{1,x} + F_{2,x} + F_{3,x} = F_1\cos(\theta_1) + F_2\cos(\theta_2) + F_{3,x} = ma_x = 0$$

$$F_{3,x} = -F_1\cos(\theta_1) - F_2\cos(\theta_2) = -(100\text{ N})\cos(60°) - (200\text{ N})\cos(37°) = \boxed{-210\text{ N}}$$

$$\sum F_y = -F_{1,y} + F_{2,y} + F_{3,y} = -F_1\sin(\theta_1) + F_2\sin(\theta_2) + F_{3,y} = ma_y = 0$$

$$F_{3,y} = F_1\sin(\theta_1) - F_2\sin(\theta_2) = (100\text{ N})\sin(60°) - (200\text{ N})\sin(37°) = \boxed{-34\text{ N}}$$

Part b)

$$F_3 = \sqrt{F_{3,x}^2 + F_{3,y}^2} = \sqrt{(-210\text{ N})^2 + (-34\text{ N})^2} = 213\text{ N}$$

$$\theta_3 = \arctan\left(\frac{F_{3,y}}{F_{3,x}}\right) = \arctan\left(\frac{-34\text{ N}}{-210\text{ N}}\right) = 189°$$

Figure 4-4 Problem 61

Part c)

$$\sum F_x = F_{1,x} + F_{2,x} = F_1\cos(\theta_1) + F_2\cos(\theta_2) = ma_x$$

$$F_1\cos(\theta_1) + F_2\cos(\theta_2) = (100\ \text{N})\cos(60°) + (150\ \text{N})\cos(37°) = 170\ \text{N} = ma_x$$

$$\sum F_\perp = -F_{1,\perp} + F_{2,\perp} = -F_1\sin(\theta_1) + F_2\sin(\theta_2) = ma_\perp$$

$$\sum F_y = -F_{1,y} + F_{2,y} = -F_1\sin(\theta_1) + F_2\sin(\theta_2) = ma_y$$

$$-F_1\sin(\theta_1) + F_2\sin(\theta_2) = -(100\ \text{N})\sin(60°) + (150\ \text{N})\sin(37°) = ma_y$$

$$\theta = \arctan\left(\frac{3.7\ \text{N}}{170\ \text{N}}\right) = \boxed{1.2°}$$

Part d)

$$a_x = a\cos(\theta) = \left(10\frac{\text{m}}{\text{s}^2}\right)\cos(1.2°) = 10\frac{\text{m}}{\text{s}^2}$$

$$m = \frac{170\ \text{N}}{a_x} = \frac{170\ \text{N}}{10\dfrac{\text{m}}{\text{s}^2}} = \boxed{17\ \text{kg}}$$

REFLECT

Players 1 and 2 pull toward $+x$, so Player 3 must pull toward $-x$ if the box is to remain stationary. If Player 3 stops pulling, the box must accelerate toward $+x$.

Get Help: P'Cast 4.2 – Three-Team Tug-of-War

4.65

SET UP

Two ropes support a 100-kg streetlight. One rope pulls up and to the left at an angle of 40 degrees above the horizontal with a tension T_1. The other rope pulls up and to the right at an angle of 40 degrees above the horizontal with a tension T_2. The forces acting on the streetlight are the tension in the first rope, the tension in the second rope, and the force of gravity pointing straight down. The streetlight is at rest, which means its acceleration is equal to zero in all directions. We can solve for the tension in each rope by solving Newton's second law in component form. The tension in each rope is equal to the magnitude T_1 or T_2.

Figure 4-5 Problem 65

SOLVE

$$\sum F_x = T_{1,x} + T_{2,x} = -T_1\cos(40°) + T_2\cos(40°) = 0$$

$$T_1 = T_2 \equiv T$$

$$\sum F_y = T_{1,y} + T_{2,y} + F_{g,y} = T_1\sin(40°) + T_2\sin(40°) - mg = 0$$

$$2T\sin(40°) = mg$$

$$T = \frac{mg}{2\sin(40°)} = \frac{(100\ \text{kg})\left(9.8\dfrac{\text{m}}{\text{s}^2}\right)}{2\sin(40°)} = \boxed{762\ \text{N}}$$

REFLECT

Because the ropes are symmetric, they must have the same tension.

Get Help: Interactive Example – Mass on Two Strings

4.69

SET UP

A train is made up of a locomotive and 10 identical freight cars, each of mass M. The magnitude of the train's acceleration is 2 m/s². The magnitude of the force between the locomotive and the first car is 100,000 N. If we treat all 10 freight cars as the "train," the force between the locomotive and the "train" is equal to the force between the locomotive and the first car. We can draw a free-body diagram of the "train" in order to calculate the mass of one of the cars. We can then draw the free-body diagram of the tenth car and apply Newton's second law in order to calculate the magnitude of the force between the ninth and tenth cars. We'll define positive x as pointing to the right in the above figure.

Figure 4-6 Problem 69

SOLVE

Free-body diagram of the "train":

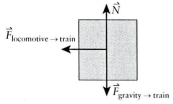

Figure 4-7 Problem 69

Newton's second law for the "train":

$$\sum F_x = -F_{\text{locomotive}\rightarrow\text{train}} = (10M)a_x$$

$$M = \frac{-F_{\text{locomotive}\rightarrow\text{train}}}{10a_x}$$

Free-body diagram of car 10:

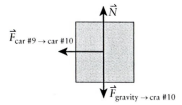

Figure 4-8 Problem 69

Newton's second law for car 10:

$$\sum F_x = -F_{\text{car } 9\rightarrow\text{car } 10} = Ma_x = \left(\frac{-F_{\text{locomotive}\rightarrow\text{train}}}{10a_x}\right)a_x$$

$$F_{\text{car } 9\rightarrow\text{car } 10} = \frac{F_{\text{locomotive}\rightarrow\text{train}}}{10} = \frac{100,000 \text{ N}}{10} = \boxed{10,000 \text{ N}}$$

REFLECT

Rather than treating all 10 cars as one object called the "train," we could have drawn 10 different free-body diagrams—one for each freight car. The force of car 1 on car 2 is equal in magnitude and opposite in direction to the force of car 2 on car 1. Applying Newton's second and third laws for each car in the series will give you the same answer as above.

4.73

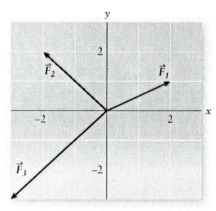

Figure 4-9 Problem 73

SET UP

We are shown three forces acting on an object and asked to calculate the acceleration of the object. The acceleration of the object is related to the net force acting on the object through Newton's second law. We can read the components of each force directly from the graph, add them with the correct sign, and solve for the components of the acceleration.

SOLVE

x component:

$$\sum F_x = F_{1,x} + F_{2,x} + F_{3,x} = (2\text{ N}) - (2\text{ N}) - (3\text{ N}) = -3\text{ N} = (2\text{ kg})a_x$$

$$a_x = -1.5\frac{\text{m}}{\text{s}^2}$$

y component:

$$\sum F_y = F_{1,y} + F_{2,y} + F_{3,y} = (1\text{ N}) + (2\text{ N}) - (3\text{ N}) = 0\text{ N} = (2\text{ kg})a_y$$

$$a_y = 0$$

$$\boxed{\vec{a} = -\left(1.5\frac{\text{m}}{\text{s}^2}\right)\hat{x}}$$

REFLECT

Remember, when calculating the net force, to split it up into components and solve each equation separately.

Get Help: P'Cast 4.2 – Three-Team Tug-of-War

4.77

SET UP

A 60.0-kg person is riding in a car. While the car uniformly accelerates from 0 to 28 m/s, the person feels a horizontal force of 400 N; this is the only force acting on the person in the horizontal direction. We can use Newton's second law to calculate the acceleration and then use the definition of constant acceleration to calculate the time it takes the car to reach 28 m/s.

SOLVE

Newton's second law for the person:

$$\sum F_x = F_{\text{car}\rightarrow\text{person}} = ma_x$$

$$a_x = \frac{F_{\text{car}\rightarrow\text{person}}}{m}$$

Calculating the time:

$$a_x = \frac{\Delta v}{\Delta t}$$

$$\Delta t = \frac{\Delta v}{a_x} = \frac{\Delta v}{\left(\dfrac{F_{car \to person}}{m}\right)} = \frac{m\Delta v}{F_{car \to person}} = \frac{(60.0 \text{ kg})\left(\left(28\dfrac{m}{s}\right) - 0\right)}{400 \text{ N}} = \boxed{4.2 \text{ s}}$$

REFLECT

The acceleration of the car is constant, so we could use the constant acceleration equations to calculate other quantities, for example, the distance the car travels.

4.81

SET UP

Two blocks, resting on two different inclined planes, are attached by a string. The block on the left has a mass $M_1 = 6$ kg and rests on an incline at 60 degrees. The block on the right has a mass M_2 and rests on an incline at 25 degrees. We need to find the value of M_2 such that both blocks remain at rest. We'll use two different but related coordinate systems for the

Figure 4-10 Problem 81

two blocks. For both blocks the axes will be parallel and perpendicular to the inclined plane. For block M_1 up the ramp and out of the ramp are positive, while down the ramp and out of the ramp are positive for M_2. We can draw free-body diagrams and apply Newton's second law in component form for both blocks. The tension acting on each block will be identical in magnitude. We are only interested in the forces and acceleration acting parallel to the planes because the blocks do not leave the plane; the acceleration in the perpendicular direction is always zero. Both blocks are at rest, which means the acceleration of each block is zero and we can calculate M_2.

SOLVE

Free-body diagram of M_1:

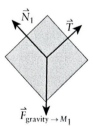

Figure 4-11 Problem 81

Newton's second law for M_1:

$$\sum F_\parallel = T - F_{gravity \to M_1, \parallel} = T - M_1 g \sin(60°) = M_1 a_\parallel = 0$$

$$T = M_1 g \sin(60°)$$

Free-body diagram for M_2:

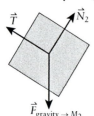

$\vec{F}_{\text{gravity} \rightarrow M_2}$ **Figure 4-12** Problem 81

Newton's second law for M_2:

$$\sum F_{\parallel} = -T + F_{\text{gravity} \rightarrow M_2,\ \parallel} = -T + M_2 g \sin(25°) = M_2 a_{\parallel} = 0$$

$$-M_1 g \sin(60°) + M_2 g \sin(25°) = 0$$

$$M_2 = \frac{M_1 \sin(60°)}{\sin(25°)} = \frac{(6 \text{ kg}) \sin(60°)}{\sin(25°)} = \boxed{12.3 \text{ kg}}$$

REFLECT

The angle on the left is steeper than the angle on the right, so the mass on the left should be smaller than the mass on the right.

Get Help: Interactive Example – Mass on Two Strings

4.85

SET UP

A 30-kg dog stands on a scale, and both are placed in an elevator. The elevator begins to move and we are interested in calculating the reading on the scale for various accelerations of the elevator, which are equal to the accelerations of the scale and dog since they travel together as one object. The reading on the scale is equal in magnitude to the normal force acting on the dog. Gravity also acts on the dog. We can use Newton's second law to calculate N as a function of the acceleration. We'll define up to be positive y throughout our calculation.

SOLVE

Free-body diagram of the dog:

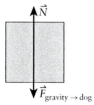

$\vec{F}_{\text{gravity} \rightarrow \text{dog}}$

Figure 4-13 Problem 85

$$\sum F_y = N - F_{\text{gravity} \rightarrow \text{dog}} = ma_y$$

$$N = F_{\text{gravity} \rightarrow \text{dog}} + ma_y = mg + ma_y = m(g + a_y)$$

Part a)

$$N = m(g + a_y) = (30 \text{ kg})\left(\left(9.8\frac{\text{m}}{\text{s}^2}\right) - \left(3.5\frac{\text{m}}{\text{s}^2}\right)\right) = \boxed{189 \text{ N}}$$

Part b)

$$N = m(g + a_y) = (30 \text{ kg})\left(\left(9.8\frac{\text{m}}{\text{s}^2}\right) + 0\right) = \boxed{294 \text{ N}}$$

Part c)

$$N = m(g + a_y) = (30 \text{ kg})\left(\left(9.8\frac{\text{m}}{\text{s}^2}\right) + \left(4.0\frac{\text{m}}{\text{s}^2}\right)\right) = \boxed{414 \text{ N}}$$

REFLECT

When an elevator starts to move downward from rest, you feel lighter; when an elevator travels at a constant speed you feel "normal"; and, when an elevator moves upward from rest, you feel heavier. All of these observations from everyday life correspond to our calculation with respect to the dog and the scale.

Get Help: Picture it – Apparent Weight

4.89

SET UP

A froghopper has a mass of 12.3 mg and can flex its legs 2.0 mm in order to jump an additional 426 mm to a height of 428 mm above the ground. We are interested in the portion of the jump while the insect is still on the ground. We can assume that the acceleration of the insect is constant during this phase. In order to calculate the acceleration of the insect and the length of time during this phase of the jump we first need to calculate the speed with which the froghopper leaves the ground. While the insect is in the air, it is only under the influence of gravity, so we can use the constant acceleration equations and the height of the jump to calculate the takeoff speed. Once we have this value, we know the froghopper accelerated from rest through a distance of 2.0 mm to reach this takeoff speed. We can directly calculate the time because we are assuming the acceleration is constant. The forces acting on the insect while it is on the ground are the force of the ground on the froghopper pointing up (that is, the normal force) and the force of gravity pointing down. After defining a coordinate system where positive y points upward, we can calculate the magnitude of the normal force using Newton's second law.

SOLVE

Part a)

Takeoff speed:

$$v^2 - v_0^2 = 0^2 - v_0^2 = 2a_y(\Delta y)$$

$$v_0 = \sqrt{-2a_y(\Delta y)} = \sqrt{-2\left(-9.8\frac{\text{m}}{\text{s}^2}\right)(0.426 \text{ m})} = 2.89\frac{\text{m}}{\text{s}}$$

Acceleration necessary to attain that speed from rest:

$$v^2 - v_0^2 = 2a_y(\Delta y)$$

$$a_y = \frac{v^2 - v_0^2}{2(\Delta y)} = \frac{\left(2.89\frac{m}{s}\right)^2 - 0}{2(0.0020 \text{ m})} = \boxed{2087\frac{m}{s^2}}$$

Time:

$$a_y = \frac{\Delta v}{\Delta t}$$

$$\Delta t = \frac{\Delta v}{a_y} = \frac{2.89\frac{m}{s}}{2087\frac{m}{s^2}} = \boxed{0.00138 \text{ s} = 1.38 \text{ ms}}$$

Part b)

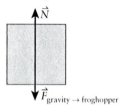

Figure 4-14 Problem 89

Part c)

$$\sum F_y = N - F_{gravity \to froghopper} = N - mg = ma_y$$

$$N = mg + ma_y = m(g + a_y) = (12.3 \times 10^{-6} \text{ kg})\left(\left(9.8\frac{m}{s^2}\right) + \left(2087\frac{m}{s^2}\right)\right)$$

$$= \boxed{0.0258 \text{ N} = 25.8 \text{ mN}}$$

This force is $\boxed{214 \text{ times larger}}$ than the froghopper's weight.

REFLECT

A froghopper is around 5mm long, so a height of 426 mm is about 85 times its length.

4.93

SET UP

Sue and Paul are attached by a rope while climbing a glacier with a 45-degree slope. Suddenly, Sue ($m_{Sue} = 66$ kg) falls into a crevasse and falls 2 m in 10 s from rest. We'll use two different but related coordinate systems for the two people. For Sue, positive y will point upward. For Paul, the axes will be parallel and perpendicular to the inclined plane, where up the ramp and out of the ramp are positive. Tension from the rope pulling up and gravity pulling down are the only forces acting on Sue. Assuming

Figure 4-15 Problem 93

her acceleration is constant, we can use the constant acceleration equations and Newton's second law to calculate the magnitude of the tension. Since Paul and Sue are tethered to one another, the magnitudes of their accelerations are equal. The tension in the rope and gravity are the only forces acting on Paul that have components that are parallel to the face of the glacier. We can then solve the parallel component of Newton's second law for Paul's mass.

SOLVE
Part a)
Sue's acceleration:

$$\Delta y = v_0 t + \frac{1}{2}a_y t^2 = 0 + \frac{1}{2}a_y t^2$$

$$a_y = \frac{2(\Delta y)}{t^2} = \frac{2(-2 \text{ m})}{(10 \text{ s})^2} = -0.04\frac{\text{m}}{\text{s}^2}$$

Free-body diagram of Sue:

Figure 4-16 Problem 93

Newton's second law for Sue:

$$\sum F_y = T - F_{\text{gravity}\rightarrow\text{Sue}} = T - m_{\text{Sue}}g = m_{\text{Sue}}a_y$$

$$T = m_{\text{Sue}}(g + a_y) = (66 \text{ kg})\left(\left(9.8\frac{\text{m}}{\text{s}^2}\right) + \left(-0.04\frac{\text{m}}{\text{s}^2}\right)\right) = \boxed{644 \text{ N}}$$

Part b)
Free-body diagram of Paul:

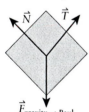

Figure 4-17 Problem 93

Newton's second law for Paul:

$$\sum F_\parallel = T - F_{\text{gravity}\rightarrow\text{Paul}, \parallel} = T - m_{\text{Paul}}g\sin(45°) = m_{\text{Paul}}a_\parallel$$

$$m_{\text{Paul}} = \frac{T}{a_\parallel + g\sin(45°)} = \frac{(644 \text{ N})}{\left(0.04\frac{\text{m}}{\text{s}^2}\right) + \left(9.8\frac{\text{m}}{\text{s}^2}\right)\sin(45°)} = \boxed{92.4 \text{ kg}}$$

REFLECT
The tension in the rope (644 N) is less than Sue's weight (647 N), so there is a net force on Sue acting downward, which means she'll accelerate in that direction. Paul's mass of 92 kg is around 200 lb, which is a reasonable weight for a human being.

Get Help: Interactive Example – Mass on Two Strings

4.97

SET UP

A man sits in a bosun's chair (see figure). Since we are given the combined mass of the man, chair, and bucket (m = 95.0 kg), we will treat them together as one object. There are two tension forces upward on this object (one from each end of the rope) and the force of gravity acting downward. Using a coordinate system where positive y points upward, we can solve Newton's second law for the tension in terms of the acceleration. Recall that an object traveling at a constant velocity has an acceleration of zero.

Figure 4-18 Problem 97

SOLVE

Free-body diagram of the (Man + Chair + Bucket):

Figure 4-19 Problem 97

Newton's second law for the (Man + Chair + Bucket):

$$\sum F_y = 2T - F_{\text{gravity}\rightarrow(\text{Man}+\text{Chair}+\text{Bucket})} = 2T - mg = ma_y$$

$$T = \frac{m(g + a_y)}{2}$$

Part a) At constant speed:

$$T = \frac{(95.0 \text{ kg})\left(\left(9.8\frac{\text{m}}{\text{s}^2}\right) + 0\right)}{2} = \boxed{466 \text{ N}}$$

Part b) At a constant upward acceleration of 1.5 m/s²:

$$T = \frac{(95.0 \text{ kg})\left(\left(9.8\frac{\text{m}}{\text{s}^2}\right) + \left(1.5\frac{\text{m}}{\text{s}^2}\right)\right)}{2} = \boxed{537 \text{ N}}$$

REFLECT

In order to remain stationary or travel at a constant speed, the man needs to pull with a force equal to the weight. The man should pull harder if he wants to accelerate upward, which makes sense.

4.101

SET UP

Blocks A and B are connected by a string with a tension T_1. Block C, which is attached to block A by a string with tension T_2, is hanging off of the table. Block A has a mass m_A = 2 kg, block B has a mass of m_B = 1 kg, and block C

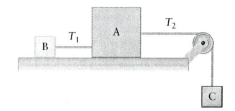

Figure 4-20 Problem 101

has a weight of 10 N (that is, $m_C = 1$ kg). We are told that blocks A and B accelerate to the right, which we'll call positive x. This means block C is falling (toward negative y). All three blocks have the same acceleration because they are tethered together. Using Newton's second law for each of the blocks, we can solve for the acceleration in terms of the known masses and weight. Once we have the acceleration, we can calculate the magnitudes of the tensions.

SOLVE

Free-body diagram of block A:

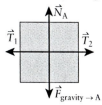

Figure 4-21 Problem 101

Free-body diagram of block B:

Figure 4-22 Problem 101

Free-body diagram of block C:

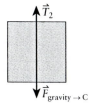

Figure 4-23 Problem 101

Newton's second law:

$$\sum F_{A,x} = -T_1 + T_2 = m_A a$$

$$\sum F_{B,x} = T_1 = m_B a$$

$$\sum F_{C,y} = T_2 - F_{gravity \to C} = m_C(-a)$$

Solving for the acceleration:

$$T_2 = F_{gravity \to C} - m_C a = (m_A + m_B)a$$

$$a = \frac{F_{gravity \to C}}{m_A + m_B + m_C}$$

Solving for the tensions:

$$T_1 = m_B a = m_B\left(\frac{F_{gravity \to C}}{m_A + m_B + m_C}\right) = (1 \text{ kg})\left(\frac{10 \text{ N}}{(2 \text{ kg}) + (1 \text{ kg}) + (1 \text{ kg})}\right) = \boxed{2.5 \text{ N}}$$

$$T_2 = (m_A + m_B)a = (m_A + m_B)\left(\frac{F_{gravity \to C}}{m_A + m_B + m_C}\right)$$

$$= ((2 \text{ kg}) + (1 \text{ kg}))\left(\frac{10 \text{ N}}{(2 \text{ kg}) + (1 \text{ kg}) + (1 \text{ kg})}\right) = \boxed{7.5 \text{ N}}$$

REFLECT

It makes sense that $T_2 > T_1$ since the second string has to pull *both* blocks A and B, while the first string is only pulling block B. We could have also treated all three blocks as one object with a combined mass of 4 kg. The only external force acting on this system is $F_{gravity \to C} = 10$ N. This would have quickly given us the acceleration for the system.

Chapter 5
Applications of Newton's Laws

Conceptual Questions

5.3 When a pickup truck accelerates forward, the contents will also accelerate forward due to the force of static friction between the box and the truck's bed.

5.7 If the water stays in the bucket, it is because the whirling is so fast that at the top of the circle, the bucket is accelerating downward faster than the acceleration of gravity. The bottom of the bucket is needed to pull the water down! The sides of the bucket prevent the water from sloshing ahead or behind with the rest of the water that is going in the same circle. The weight is always pointing straight down (whether at the top or at the bottom), but the normal force N changes direction, pointing down at the top of the swing and up at the lowest point. Thus, $\sum F_{\text{top}} = mg + N$ while $\sum F_{\text{bottom}} = mg - N$.

5.11 His acceleration decreases because the net force on him decreases. Net force is equal to his weight minus his drag force. The faster an object is moving, the larger the drag force it experiences as a function of its speed. Because drag force increases with increasing speed, net force and acceleration decrease. As the drag force approaches his weight, the acceleration approaches zero.

5.15 If you know the car's instantaneous speed and the radius of curvature, you can use the formula $a_c = \dfrac{v^2}{r}$. A more direct alternative is to use an accelerometer—a circular glass tube with a ball in it. If the ball settles at angle α from the bottom, tan α = (centripetal acceleration)/(acceleration of gravity).

5.17 Banking the curve allows the normal force to help push the car into the turn. The additional normal force also increases the maximum frictional force.

Multiple-Choice Questions

5.21 **A** (slide down at constant speed).
Free-body diagram of a mass sliding down a ramp:

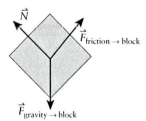

Figure 5-1 Problem 21

Newton's second law:

$$\sum F_\perp = N - mg\cos(45°) = ma_\perp = 0$$

$$N = mg\cos(45°)$$

$$\sum F_\| = \mu_k N - mg\sin(45°) = ma_\|$$

$$\mu_k(mg\cos(45°)) - mg\sin(45°) = ma_\|$$

$$a_\| = \mu_k g\cos(45°) - g\sin(45°)$$

The acceleration is independent of the mass, which means the blocks will have the same motion.

Get Help: Interactive Example – Friction Fraction
P'Cast 5.3 – Up a Ramp

5.25 D (the drag force is larger than the skydiver's weight). Once the skydiver opens his parachute, he will start to slow down because there is a net force acting upward. A net force upward means that the magnitude of the drag force is larger than the magnitude of his weight.

5.29 B ($N > mg$). There is a net force acting upward because there is an acceleration in that direction when you are at the bottom of a rotating Ferris wheel.

Estimation Questions

5.33 Assume the initial speed of the runner to be 7 m/s and the slide covers 3 m. The acceleration would be $a = \dfrac{-\left(7\dfrac{m}{s}\right)^2}{2(3\ m)} = -8\dfrac{m}{s^2}$.

$$\sum F = \mu mg = ma$$

$$\mu = \frac{a}{g} = \frac{\left(8\dfrac{m}{s^2}\right)}{\left(9.8\dfrac{m}{s^2}\right)} = \boxed{0.8}$$

5.37 It takes about 2 s to fall 2 m at a 30-degree angle from vertical (the length of the slide); with these data, the coefficient of kinetic friction is 0.34.

Problems

5.41

SET UP

A 7.6-kg object rests on a level floor with a coefficient of static friction of 0.55. The minimum horizontal applied force that will cause the object to start sliding is equal to the maximum possible magnitude of the static frictional force. This will have a magnitude equal to the coefficient of static friction multiplied by the normal force acting on the object. Because the object is stationary in the vertical direction, the normal force is equal in magnitude to the weight.

SOLVE

$$F_{app} = F_{sf, max} = \mu_s N = \mu_s mg = (0.55)(7.6 \text{ kg})\left(9.8\frac{\text{m}}{\text{s}^2}\right) = \boxed{41 \text{ N}}$$

REFLECT

Usually the magnitude of static friction involves an inequality. We can use an equal sign here because we're looking for the maximum static frictional force.

5.45

SET UP

A horizontal force $F_{app} = 10$ N is applied to a stationary 2-kg block. The coefficient of static friction between the block and the floor is 0.75. In order to determine the motion of the block, we first need to determine which friction—static or kinetic—needs to be considered. Newton's second law in the vertical direction will give us the magnitude of the normal force, which is used in

Figure 5-2 Problem 45

calculating the magnitudes of the frictional forces. We should first calculate the magnitude of the static frictional force. If it is larger than 10 N, the block won't move; if it's less than 10 N, then we calculate the magnitude of kinetic friction and use that to determine the acceleration in the horizontal direction.

SOLVE

Free-body diagram of block:

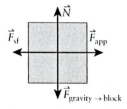

Figure 5-3 Problem 45

Newton's second law:

$$\sum F_y = N - F_{gravity \to block} = N - mg = ma_y = 0$$

$$N = mg$$

Magnitude of static friction:

$$F_{sf} \leq \mu_s N = \mu_s mg = (0.75)(2 \text{ kg})\left(9.8 \frac{\text{m}}{\text{s}^2}\right) = 14.7 \text{ N}$$

The applied force of 10 N is not enough to overcome static friction, which means the block will remain stationary.

REFLECT
The information regarding kinetic friction is irrelevant for this problem.

5.49

SET UP
A book is pushed across a horizontal table with a force equal to one-half the book's weight. The book is traveling at a constant speed, which means its acceleration in this direction is equal to zero. The only forces acting in the horizontal direction are the applied force and kinetic friction, and they act in opposite directions. The magnitude of the kinetic frictional force is constant and equal to the product of the coefficient of kinetic friction and the normal force. Assuming that the object is not accelerating in the vertical direction, the magnitude of the normal force is equal to the object's weight. Setting the forces equal we can solve for the coefficient of kinetic friction.

SOLVE

$$\sum F_{horizontal} = F_{app} - F_{kf} = ma_{horizontal} = 0$$

$$F_{app} = F_{kf} = \mu_k N = \mu_k F_{gravity \rightarrow book}$$

$$\mu_k = \frac{F_{app}}{F_{gravity \rightarrow book}} = \frac{\left(\dfrac{F_{gravity \rightarrow book}}{2}\right)}{F_{gravity \rightarrow book}} = \boxed{0.5}$$

REFLECT
This is a reasonable value for the coefficient of kinetic friction for a book sliding across a table.

5.53

SET UP
A 12-kg block and a 5-kg block are connected by a taut string with tension T. The coefficient of static friction between the 12-kg block and the floor is $\mu_{s, 12\text{-kg}} = 0.443$ and the coefficient of static friction between the 5-kg block and the floor is

Figure 5-4 Problem 53

$\mu_{s, 5\text{-kg}} = 0.573$. An external force F pulling to the right is applied to the 5-kg block. We need to determine, using Newton's second law, the minimum F necessary to overcome static friction and accelerate the blocks toward positive x. Since the mass is positive and the x component of the acceleration will be positive when the blocks move, the net force acting on each block also needs to be positive (that is, greater than zero).

SOLVE

Free-body diagram of the 12-kg block:

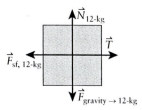

Figure 5-5 Problem 53

Free-body diagram of the 5-kg block:

Figure 5-6 Problem 53

Newton's second law, y components:

$$\sum F_{12\text{-kg}, y} = N_{12\text{-kg}} - F_{\text{gravity}\to 12\text{-kg}} = N_{12\text{-kg}} - m_{12\text{-kg}}g = m_{12\text{-kg}}a_y = 0$$

$$N_{12\text{-kg}} = m_{12\text{-kg}}g$$

$$\sum F_{5\text{-kg}, y} = N_{5\text{-kg}} - F_{\text{gravity}\to 5\text{-kg}} = N_{5\text{-kg}} - m_{5\text{-kg}}g = m_{5\text{-kg}}a_y = 0$$

$$N_{5\text{-kg}} = m_{5\text{-kg}}g$$

Newton's second law, x components:

$$\sum F_{12\text{-kg}, x} = T - F_{\text{sf}\to 12\text{-kg}} = T - \mu_{\text{s, 12-kg}}N_{12\text{-kg}} = T - \mu_{\text{s, 12-kg}}m_{12\text{-kg}}g = m_{12\text{-kg}}a_x$$

$$\sum F_{5\text{-kg}, x} = F - T - F_{\text{sf}\to 5\text{-kg}} = F - T - \mu_{\text{s, 5-kg}}N_{5\text{-kg}} = F - T - \mu_{\text{s, 5-kg}}m_{5\text{-kg}}g = m_{5\text{-kg}}a_x$$

Adding the x-component equations together:

$$F - g(\mu_{\text{s, 12-kg}}m_{12\text{-kg}} + \mu_{\text{s, 5-kg}}m_{5\text{-kg}}) = (m_{5\text{-kg}} + m_{12\text{-kg}})a_x$$

Setting up and solving the inequality:

$$F - g(\mu_{\text{s, 12-kg}}m_{12\text{-kg}} + \mu_{\text{s, 5-kg}}m_{5\text{-kg}}) = (m_{5\text{-kg}} + m_{12\text{-kg}})a_x \geq 0$$

$$F \geq g(\mu_{\text{s, 12-kg}}m_{12\text{-kg}} + \mu_{\text{s, 5-kg}}m_{5\text{-kg}}) = \left(9.8\frac{\text{m}}{\text{s}^2}\right)((0.443)(12 \text{ kg}) + (0.573)(5 \text{ kg})) = \boxed{80.2 \text{ N}}$$

REFLECT

The maximum value of static friction is when $F_{\text{sf}} = \mu_s N$. When the magnitude of the applied force is larger than this value, the system will start to accelerate.

5.57

SET UP

Two blocks are connected by a string. Block 1 is hanging over the edge of a ramp. Block 2 ($m_2 = 8$ kg) is resting on a 28-degree slope. The coefficient of kinetic friction between block 2 and the ramp is 0.22. We are told block 2 is sliding down the ramp at a constant speed, which means its acceleration is constant in all directions. Because the two blocks are connected, block 1 is also traveling at a constant speed. We can use Newton's second law to calculate the

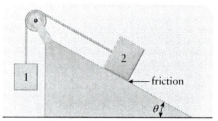

Figure 5-7 Problems 57

mass of block 1. We will use an axis that points along the string, starting from block 1, and down the ramp will be positive.

SOLVE

Free-body diagram of block 1:

Figure 5-8 Problem 57

Free-body diagram of block 2:

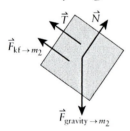

Figure 5-9 Problem 57

Newton's second law:

$$\sum F_{1,\,y} = T - F_{\text{gravity}\to 1} = T - m_1 g = m_1 a = 0$$

$$T = m_1 g$$

$$\sum F_{2,\,\perp} = N - F_{\text{gravity}\to 2,\,\perp} = N - m_2 g \cos(28°) = m_2 a_\perp = 0$$

$$N = m_2 g \cos(28°)$$

$$\sum F_{2,\,\parallel} = F_{\text{gravity}\to 2,\,\parallel} - T - F_{\text{kf}} = m_2 g \sin(28°) - T - \mu_k N$$

$$= m_2 g \sin(28°) - T - \mu_k (m_2 g \cos(28°))$$

$$= m_2 g \sin(28°) - m_1 g - \mu_k (m_2 g \cos(28°)) = 0$$

$$m_1 = m_2(\sin(28°) - \mu_k \cos(28°)) = (8 \text{ kg})(\sin(28°) - (0.22)\cos(28°)) = \boxed{2.2 \text{ kg}}$$

REFLECT

We are told block 2 is sliding down the ramp, so we should expect the mass of block 1 to be less than 8 kg; our answer makes sense.

5.61

SET UP

A 5-kg object is traveling at a terminal speed of 50 m/s. The magnitude of the drag force it experiences is equal to $F_{\text{drag}} = bv^{2.5}$. We can set the drag force equal to the weight of the object in order to calculate the value of b.

SOLVE

$$F_{\text{gravity} \to \text{object}} = F_{\text{drag}}$$

$$mg = bv^n$$

$$b = \frac{mg}{v^n} = \frac{(5 \text{ kg})\left(9.8\frac{\text{m}}{\text{s}^2}\right)}{\left(50\frac{\text{m}}{\text{s}}\right)^{2.5}} = \boxed{0.00277}$$

REFLECT

The SI units for b in this case are $\text{kg} \cdot \left(\dfrac{\text{s}}{\text{m}^3}\right)^{0.5}$.

5.65

SET UP

A 1500-kg truck is driving around an unbanked curve at a speed of 20 m/s. The maximum frictional force between the road and the tires is 8000 N. We can use Newton's second law to relate the maximum frictional force to the radius of the curvature by realizing the car is undergoing uniform circular motion and that static friction is the force responsible for the car traveling around the circle.

SOLVE

Free-body diagram of the truck:

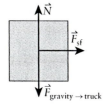

Figure 5-10 Problem 65

Newton's second law:

$$\sum F_y = N - F_{\text{gravity} \to \text{truck}} = N - mg = ma_y = 0$$

$$N = mg$$

$$\sum F_x = F_{sf} = ma_x = m\left(\frac{v^2}{R}\right)$$

$$R = \frac{mv^2}{F_{sf}} = \frac{(1500 \text{ kg})\left(20\frac{\text{m}}{\text{s}}\right)^2}{8000 \text{ N}} = \boxed{75 \text{ m}}$$

REFLECT

Although it may seem large, this is a reasonable radius for a curve on a highway. The curves on a highway are usually gentle because cars are traveling at high speeds.

5.69

SET UP

A centrifuge of radius $R = 0.1$ m spins a 1-g sample at 1200 rev/min. The net force causing it to spin is equal to the mass of the sample multiplied by the centripetal acceleration of the sample.

SOLVE

$$F = ma = \frac{mv^2}{R} = \frac{(10^{-3} \text{ kg})\left(\dfrac{1200 \text{ rev}}{\text{min}} \times \dfrac{1 \text{ min}}{60 \text{ s}} \times \dfrac{2\pi(0.1 \text{ m})}{1 \text{ rev}}\right)^2}{0.1 \text{ m}} = \boxed{1.58 \text{ N}}$$

REFLECT

This force is about 100 times larger than the weight of the sample.

5.73

SET UP

A 0.75-kg tetherball is attached to a 1.25-m rope and is spun around a pole. The rope makes an angle of 35 degrees with the vertical. Although the ball is traveling in a circle in the horizontal plane, the ball is at rest with respect to the vertical direction. We can use Newton's second law to calculate the tension in the rope.

Figure 5-11 Problem 73

SOLVE

Free-body diagram of the ball:

Figure 5-12 Problem 73

Newton's second law:

$$\sum F_y = T_y - F_{gravity} = T\cos(35°) - mg = ma_y = 0$$

$$T = \frac{mg}{\cos(35°)} = \frac{(0.75 \text{ kg})\left(9.8\frac{\text{m}}{\text{s}^2}\right)}{\cos(35°)} = \boxed{8.97 \text{ N}}$$

REFLECT

We didn't need to use the information regarding the ball's speed in order to find the tension.

Get Help: Interactive Example – Tetherball

5.77

SET UP

The coefficient of static friction between a rubber tire and dry pavement is around 0.80. We can use Newton's second law in order to calculate the acceleration of the car and then the time required for the car to accelerate from rest to 26.8 m/s. We are told that the engine supplies power to only two of the four wheels, which means we need to divide the maximum force due to static friction by two. Friction opposes relative *slipping* motion between two objects but can still cause the car to accelerate forward.

SOLVE

Part a)
Free-body diagram of the car:

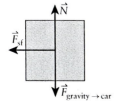

Figure 5-13 Problem 77

Newton's second law:

$$\sum F_y = N - F_{\text{gravity}\rightarrow\text{car}} = N - mg = ma_y = 0$$

$$N = mg$$

$$\sum F_x = F_{\text{sf}} = \mu_s\left(\frac{N}{2}\right) = \mu_s\left(\frac{mg}{2}\right) = ma_x$$

$$a_x = \frac{\mu_s g}{2} = \frac{(0.80)\left(9.8\frac{\text{m}}{\text{s}^2}\right)}{2} = 3.92\frac{\text{m}}{\text{s}^2}$$

Calculating the time:

$$a_x = \frac{\Delta v}{\Delta t}$$

$$\Delta t = \frac{\Delta v}{a_x} = \frac{\left(26.8\frac{\text{m}}{\text{s}}\right)}{\left(3.92\frac{\text{m}}{\text{s}^2}\right)} = \boxed{6.84 \text{ s}}$$

Part b) Static friction opposes the relative slipping motion between two surfaces. The bottom of the tire tends to slip backward relative to the wet pavement, so the frictional force opposes the backward motion and, hence, is forward.

REFLECT
On dry pavement, the car's tires do not slip relative to the surface of the road. An example of when the tires *do* slip would be on an icy road or on rain-soaked pavement.

5.81

SET UP
A 2.5-kg package slides down a 20-degree incline with an initial speed of 2 m/s. The package slides a total distance of 12.0 m. We can calculate the constant acceleration of the package assuming it comes to rest at the bottom of the ramp. Using Newton's second law, we can relate this acceleration to the net force on the package and then calculate the coefficient of kinetic friction between the package and the ramp.

SOLVE
Finding the acceleration:

$$v^2 - v_0^2 = 2a_{\parallel}(\Delta x)$$

$$a_{\parallel} = \frac{v^2 - v_0^2}{2(\Delta x)} = \frac{0 - \left(2\frac{m}{s}\right)^2}{2(12.0\ m)} = -0.167\frac{m}{s^2}$$

Free-body diagram of the package:

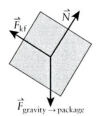

Figure 5-14 Problem 81

Newton's second law:

$$\sum F_{\perp} = N - F_{gravity \rightarrow package,\ \perp} = N - mg\cos(20°) = ma_{\perp} = 0$$

$$N = mg\cos(20°)$$

$$\sum F_{\parallel} = F_{gravity \rightarrow package,\ \parallel} - F_{kf} = mg\sin(20°) - \mu_k N = mg\sin(20°) - \mu_k mg\cos(20°) = ma_{\parallel}$$

$$\mu_k = \frac{g\sin(20°) - a_{\parallel}}{g\cos(20°)} = \frac{\left(9.8\frac{m}{s^2}\right)\sin(20°) - \left(-0.167\frac{m}{s^2}\right)}{\left(9.8\frac{m}{s^2}\right)\cos(20°)} = \boxed{0.382}$$

REFLECT

We can look at the limiting cases in order to double-check our algebraic solution. As the angle of the ramp approaches 0 degrees, the magnitude of kinetic friction should be smaller if the block comes to rest at the end of the ramp. As the angle of the ramp approaches 90 degrees, then the coefficient of kinetic friction will need to be extremely large in order to attain the same acceleration.

5.85

SET UP

In a weird parallel universe, an object moving in a liquid experiences a drag force inversely proportional to its speed: $F_{drag} = -bv^{-1}$. An object of mass m is initially moving with a speed of v_0. Assuming the drag force is the only force acting along the direction of the object's motion, we can use the differential form of Newton's second law to solve for the speed as a function of time. This is a separable differential equation with the initial condition of $v(0) = v_0$. We are told that a specific object enters the liquid with an initial speed of 10 m/s, and after 9 s its speed is 8 m/s. We can use our algebraic expression for $v(t)$ from part (a) along with these data to calculate the ratio b/m and then the time necessary for the object to come to a complete stop.

SOLVE

Part a)

$$\sum F = F_{drag} = -bv^{-1} = ma = m\frac{dv}{dt}$$

$$\frac{dv}{dt} = -\frac{b}{mv}$$

$$v\,dv = -\frac{b}{m}dt$$

$$\int v\,dv = \int -\frac{b}{m}dt$$

$$\frac{1}{2}v^2 = -\frac{b}{m}t + C$$

$$v(t) = \sqrt{-\frac{2b}{m}t + C}$$

Applying the initial condition:

$$v(t = 0) = \sqrt{C} = v_0$$

$$C = v_0^2$$

$$\boxed{v(t) = \sqrt{-\frac{2b}{m}t + v_0^2}}$$

Part b)

$$v(9\text{ s}) = 8\frac{\text{m}}{\text{s}} = \sqrt{-\frac{2b}{m}(9\text{ s}) + \left(10\frac{\text{m}}{\text{s}}\right)^2}$$

$$\frac{b}{m} = 2\frac{\text{m}^2}{\text{s}^2}$$

$$v(t) = 0 = \sqrt{-\frac{2b}{m}t + v_0^2}$$

$$0 = -\frac{2b}{m}t + v_0^2$$

$$t = \frac{v_0^2}{2\left(\frac{b}{m}\right)} = \frac{\left(10\frac{\text{m}}{\text{s}}\right)^2}{2\left(2\frac{\text{m}^2}{\text{s}^3}\right)} = \boxed{25\text{ s}}$$

REFLECT
Because the drag force is inversely proportional to the speed, the largest force will be when the object is moving most slowly.

5.89

SET UP
A mass M is attached to a string of length L and is rotated. The angle the string makes with the vertical is θ. The speed of the mass is N revolutions per second. In one revolution the mass sweeps out a circle of radius $R = L\sin(\theta)$. We can use Newton's second law to solve for the observed value of θ. The only forces acting on the mass are tension from the string and gravity. The mass is stationary in the y direction and undergoing centripetal motion in the x direction. Keep in mind that we are only allowed to have M, L, N, and physical constants in our final answer.

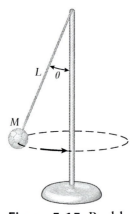

Figure 5-15 Problem 89

SOLVE
Free-body diagram of M:

Figure 5-16 Problem 89

Newton's second law:

$$\sum F_x = T_x = T\sin(\theta) = M\left(\frac{v^2}{R}\right) = M\left(\frac{(N(2\pi R))^2}{R}\right) = M(4\pi^2 N^2 R) = 4\pi^2 N^2 M(L\sin(\theta))$$

$$T = 4\pi^2 N^2 ML$$

$$\sum F_y = T_y - F_{\text{gravity}\to M} = T\cos(\theta) - Mg = Ma_y = 0$$

$$T\cos(\theta) = 4\pi^2 N^2 ML\cos(\theta) = Mg$$

$$\cos(\theta) = \frac{g}{4\pi^2 N^2 L}$$

$$\boxed{\theta = \arccos\left(\frac{g}{4\pi^2 N^2 L}\right)}$$

REFLECT

If we observe that the mass makes a smaller angle with the vertical, we would assume the rotation speed has increased. It also makes sense that a smaller angle could result from a longer string (that is, increased L).

Get Help: Interactive Example – Tetherball

5.94

SET UP

A small object moving through a vat of oil experiences a drag force of $F_{\text{drag}} = -bv^{1/2}$. Since we are ignoring the force of the gravity, the drag force is the only force acting on the object. Therefore, it is equal to its mass times its acceleration. The acceleration is the first derivative of velocity with respect to time, which means we have a differential equation. We can solve this separable differential equation for $v(t)$ by sequestering like terms and integrating. The initial condition that $v(t = 0) = v_0$ allows us to find the constant of integration.

SOLVE

$$\sum F = ma = m\frac{dv}{dt} = F_{\text{drag}} = -bv^{1/2}$$

$$\frac{dv}{v^{1/2}} = -\frac{b}{m}dt$$

$$\int \frac{dv}{v^{1/2}} = \int -\frac{b}{m}dt$$

$$2v^{1/2} = -\frac{b}{m}t + C$$

$$v^{1/2} = -\frac{b}{2m}t + C$$

Applying the initial conditions ($v(t = 0) = v_0$):

$$v_0^{1/2} = C = \sqrt{v_0}$$

$$\boxed{v(t) = \left(-\frac{b}{2m}t + \sqrt{v_0}\right)^2}$$

REFLECT

The dimensions of our expression are valid since b has SI units of $\dfrac{\text{kg} \cdot \text{m}^{1/2}}{\text{s}^{3/2}}$. As time goes on, the speed of the object decreases as expected.

Chapter 6
Work and Energy

Conceptual Questions

6.3 Stepping on the log would require raising the body, which would require doing work on it. Most of the work would not be recovered when coming down off the log.

Get Help: P'Cast 6.1 – Lifting a Water Bottle

6.7 If the surface is moving, yes. Otherwise, the distance over which the force is applied is zero, so the work will be zero.

6.11 The snowboarder walks to the top of the mountain, which requires converting stored chemical energy into heat and kinetic energy, which in turn is converted into gravitational potential energy and more heat (if the snowboarder rides a lift, the source will also include wherever the lift gets its energy). Then, riding down, the gravitational potential energy and some chemical energy will be converted into kinetic energy, which is in turn dissipated into heat.

6.15 (a) When it reaches the bottom. (b) When it reaches the bottom (but it's been close to that speed for a while).

6.19 In the special case of constant force, the integral is just a multiplication problem, so you do not need to be aware of calculus to be able to determine it. If the force is not constant and/or the path is complicated, you will need to calculate an integral.

Multiple-Choice Questions

6.23 C (decreases). As θ increases to 90 degrees, $\cos(\theta)$ approaches zero.

Get Help: P'Cast 6.2 – Pushing a Car

6.27 C (They have the same kinetic energy.) Each twin starts with the same gravitational potential energy and falls the same distance.

Estimation Questions

6.31 A standard estimate of the power of hard jogging is 7–8 kcal/kg/hr. For a runner who is 80 kg and who runs for 25 min, the energy is roughly 1 MJ.

6.35 Pumas have been seen to leap 5.4 m. If the puma's average mass is 62 kg, the maximum kinetic energy is about 300 J.

Problems

6.39

SET UP

A weightlifter lifts 446 kg a distance of 2.0 m. If the mass is moving at constant acceleration, then the magnitude of the force of the weightlifter on the mass is equal to the weight of the mass. Both of these forces are constant. The force of the weightlifter on the mass is parallel to the displacement of the mass.

SOLVE

$$W_{\text{weightlifter}} = F_{\text{weightlifter}}d\cos(0°) = mgd = (446 \text{ kg})\left(9.8\frac{\text{m}}{\text{s}^2}\right)(2.0 \text{ m}) = \boxed{8700 \text{ J}}$$

REFLECT

We are only allowed two significant figures in our final answer since d has two significant figures. The force and the displacement are in the same direction, so we expect W to be positive.

Get Help: P'Cast 6.1 – Lifting a Water Bottle

6.43

$\mu = 0.54$

3 m

40°

Figure 6-1 Problem 43

SET UP

A 150-kg crate is moving down a 40-degree incline. The coefficient of kinetic friction between the crate and the ramp is 0.54. A museum curator is pushing against the crate with enough force that the crate does not accelerate. We can use Newton's second law to determine the magnitude of the curator's force and the magnitude of kinetic friction. Once we know the magnitudes and directions of all of the forces acting on the crate, we need to determine the angle between each force and the displacement of the crate before calculating the work done by each force.

SOLVE

Free-body diagram of the crate:

$\vec{F}_{\text{curator}}$ $\vec{N}$

$\vec{F}_{\text{kf}}$

$\vec{F}_{\text{gravity} \rightarrow \text{crate}}$ **Figure 6-2** Problem 43

Part a)

$$W_{\text{gravity}} = F_{\text{gravity}}d\cos(50°) = mgd\cos(50°) = (150 \text{ kg})\left(9.8\frac{\text{m}}{\text{s}^2}\right)(3 \text{ m})\cos(50°) = \boxed{2.8 \times 10^3 \text{ J}}$$

Part b)
Newton's second law:

$$\sum F_\perp = N - F_{\text{gravity}, \perp} = N - mg\cos(40°) = ma_\perp = 0$$

$$N = mg\cos(40°)$$

$$\sum F_\parallel = F_{\text{gravity}, \parallel} - F_{\text{curator}} - F_{\text{kf}} = mg\sin(40°) - F_{\text{curator}} - \mu_k N$$

$$= mg\sin(40°) - F_{\text{curator}} - \mu_k(mg\cos(40°)) = ma_\parallel = 0$$

$$F_{\text{curator}} = mg(\sin(40°) - \mu_k\cos(40°))$$

Calculating the work:

$$W_{\text{curator}} = F_{\text{curator}}d\cos(180°) = -mg(\sin(40°) - \mu_k\cos(40°))d$$

$$= -(150 \text{ kg})\left(9.8\frac{\text{m}}{\text{s}^2}\right)(\sin(40°) - (0.54)\cos(40°))(3 \text{ m}) = \boxed{-1.0 \times 10^3 \text{ J}}$$

Part c)

$$W_{\text{kf}} = F_{\text{kf}}d\cos(180°) = -(\mu_k N)d = -\mu_k(mg\cos(40°))d$$

$$= -(0.54)(150 \text{ kg})\left(9.8\frac{\text{m}}{\text{s}^2}\right)\cos(40°)(3 \text{ m}) = \boxed{-1.8 \times 10^3 \text{ J}}$$

Part d)

$$W_N = Nd\cos(90°) = \boxed{0}$$

REFLECT
Remember that the angle used when calculating work is the angle between the force vector and the displacement vector, not the angle of the inclined plane.

Get Help: P'Cast 6.1 – Lifting a Water Bottle
P'Cast 6.2 – Pushing a Car

6.47

SET UP
A 0.25-g bumblebee is moving at a speed of 10 m/s. We can calculate its kinetic energy directly from these quantities.

SOLVE

$$K = \frac{1}{2}mv^2 = \frac{1}{2}(2.5 \times 10^{-4} \text{ kg})\left(10\frac{\text{m}}{\text{s}}\right)^2 = \boxed{1 \times 10^{-2} \text{ J}}$$

REFLECT

The speed has one significant figure, so our answer has one significant figure.

6.51

SET UP

A 2-kg block, which is initially at rest, is pushed by a force of 40 N for 22 m over a frictionless surface. This is the only force that does work on the block because the normal force and the force due to gravity are perpendicular to the displacement of the block. The net work on the block is equal to the change in its kinetic energy; this allows us to calculate the block's final speed.

Figure 6-3 Problem 51

SOLVE

$$W_{\text{push}} = F_{\text{push}}d\cos(20°) = (40 \text{ N})(22 \text{ m})\cos(20°) = 830 \text{ J}$$

$$W_{\text{net}} = \Delta K = K_f - K_i = \frac{1}{2}mv_f^2 - \frac{1}{2}mv_i^2 = \frac{1}{2}m(v_f^2 - v_i^2) = \frac{1}{2}mv_f^2 - 0$$

$$v_f = \sqrt{\frac{2W_{\text{net}}}{m}} = \sqrt{\frac{2(830 \text{ J})}{2 \text{ kg}}} = \boxed{29\frac{\text{m}}{\text{s}}}$$

REFLECT

The final speed would be larger if the force were applied at an angle closer to 0 degrees.

6.55

SET UP

A 0.15-kg baseball is initially traveling at a speed of 44 m/s when it is caught by a catcher's glove. The glove slows the ball to rest over a distance of 0.125 m. The net work on the ball is equal to the work done by the glove on the ball. The force of the glove on the ball points in the opposite direction to the ball's displacement. We can use the work–kinetic energy theorem to calculate the magnitude of the average force of the glove during the catch.

SOLVE

$$\Delta K = \frac{1}{2}mv_f^2 - \frac{1}{2}mv_i^2 = 0 - \frac{1}{2}mv_i^2$$

$$W_{\text{net}} = W_{\text{glove}\rightarrow\text{ball}} = F_{\text{glove}\rightarrow\text{ball}}d\cos(180°) = -F_{\text{glove}\rightarrow\text{ball}}d$$

Using the work–kinetic energy theorem:

$$-\frac{1}{2}mv_i^2 = -F_{\text{glove}\rightarrow\text{ball}}d$$

$$F_{\text{glove}\rightarrow\text{ball}} = \frac{mv_i^2}{2d} = \frac{(0.15 \text{ kg})\left(44\frac{\text{m}}{\text{s}}\right)^2}{2(0.125 \text{ m})} = \boxed{1200 \text{ N}}$$

The force of the glove on the hand will be slightly less than the force of the glove on the ball because the glove will compress somewhat and absorb energy.

REFLECT

The force of the glove on the ball may not necessarily be (and probably is not) constant. The average force, however, is constant.

6.59

SET UP

An object is attached to a spring with a spring constant $k = 450$ N/m. The object starts at a position 0.12 m from equilibrium and is pulled to a position 0.18 m from equilibrium. The work done by the spring on the object is related to the force of the spring on the object and the displacement. Because the force of the spring is a function of position, we need to use the integral form to calculate the work done. There will also be a negative sign introduced because the force of the spring points in the opposite direction to the displacement.

SOLVE

$$W_{\text{spring}} = -\int_{0.12}^{0.18} F_{\text{spring}} dx = -\int_{0.12}^{0.18} (kx) dx = -\left[\frac{1}{2}kx^2\right]_{0.12}^{0.18} \text{ (SI units)}$$

$$= -\frac{1}{2}\left(450\frac{\text{N}}{\text{m}}\right)((0.18 \text{ m})^2 - (0.12 \text{ m})^2) = \boxed{-4.1 \text{ J}}$$

REFLECT

The magnitude of the work done by the spring increases nonlinearly as the object is pulled farther and farther from equilibrium. For example, the magnitude of the work done in moving the object from $x = 12$ cm to $x = 18$ cm will be less than the magnitude of the work done in moving the object from $x = 18$ cm to $x = 24$ cm, even though the displacement is $+6$ cm in both cases.

Get Help: Picture It – Work by a Spring

6.63

SET UP

We are given the variable force that acts on an object that moves from the origin to (3,4). We will need to use the integral form of the work because the force is a function of the object's position. We can separate the integral into a "dx" integral from $x = 0$ to $x = 3$ and a "dy" integral from $y = 0$ to $y = 4$. All of the values are in SI units, so our final answer will have units of joules.

SOLVE

$$\vec{F} = 3x\hat{x} + 4y\hat{y}$$

$$d\vec{r} = dx\hat{x} + dy\hat{y}$$

$$\int_{(0,0)}^{(3,4)} \vec{F} \cdot d\vec{r} = \int_{(0,0)}^{(3,4)} (3x\hat{x} + 4y\hat{y}) \cdot (dx\hat{x} + dy\hat{y}) = \int_0^3 3x dx + \int_0^4 4y dy$$

$$= \left[\frac{3}{2}x^2\right]_0^3 + [2y^2]_0^4 = \frac{3}{2}(3)^2 + 2(4)^2 + 0 + 0 = \boxed{46 \text{ J}}$$

REFLECT

Because x and y are positive, the force points in the same general direction as the displacement and the work should be positive.

6.67

SET UP

A 40.0-kg boy glides down a hill on a skateboard. He ends up a vertical distance of 4.35 m below his initial location. The change in his gravitational potential energy is equal to this change in height multiplied by his weight.

SOLVE

$$\Delta U_{gravity} = mg\Delta y = (40.0 \text{ kg})\left(9.8\frac{m}{s^2}\right)(-4.35 \text{ m}) = \boxed{-1700 \text{ J} = -1.7 \text{ kJ}}$$

REFLECT

The boy's initial gravitational potential energy is converted into kinetic energy as he glides down the hill.

6.71

SET UP

A 0.0335-kg coin is initially at the ground level of a building and then carried up to a height of 630 m. The change in the coin's gravitational potential energy is equal to its weight multiplied by the difference in its vertical position.

SOLVE

$$\Delta U_{gravity} = mg\Delta y = (0.0335 \text{ kg})\left(9.8\frac{m}{s^2}\right)((630 \text{ m}) - (0 \text{ m})) = \boxed{210 \text{ J}}$$

REFLECT

Technically, the answer will be a little less than this since g decreases with height.

6.75

SET UP

A water balloon is thrown straight down with an initial speed of 12 m/s from a height of 5.0 m above the ground. We can use conservation of energy to calculate the speed of the ball right before it hits the ground. We'll define the ground to have a potential energy of zero.

SOLVE

$$U_i + K_i = U_f + K_f$$

$$mgy_i + \frac{1}{2}mv_i^2 = 0 + \frac{1}{2}mv_f^2$$

$$v_f = \sqrt{2gy_i + v_i^2} = \sqrt{2\left(9.8\frac{m}{s^2}\right)(5.0 \text{ m}) + \left(12\frac{m}{s}\right)^2} = \boxed{16\frac{m}{s}}$$

REFLECT

This is the same answer we would get if we used the constant acceleration equation $v^2 - v_0^2 = 2a_y(\Delta y)$.

Get Help: Picture It – Mechanical Energy
P'Cast 6.12 – Roller Coaster

6.79

SET UP

A pendulum is made of a ball of mass m and a string of length $L = 1.25$ m. The ball is lifted to a height such that the string makes an angle of 30 degrees with respect to the vertical. All of the ball's initial gravitational potential energy is converted to kinetic energy at the bottom of its swing. Applying conservation of mechanical energy, we can determine the speed of the ball at the bottom of its swing. We will measure the gravitational potential energy relative to the lowest part of the pendulum's swing. We can use trigonometry to relate the initial height y_i to the length L.

Figure 6-4 Problem 79

SOLVE

Trigonometry:

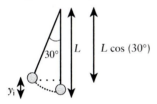

Figure 6-5 Problem 79

$$y_i = L - L\cos(30°) = L(1 - \cos(30°))$$

Conservation of energy:

$$U_i + K_i = U_f + K_f$$

$$mgy_i + 0 = 0 + \frac{1}{2}mv_f^2$$

$$gL(1 - \cos(30°)) = \frac{1}{2}v_f^2$$

$$v_f = \sqrt{2gL(1 - \cos(30°))} = \sqrt{2\left(9.8\frac{m}{s^2}\right)(1.25\ m)(1 - \cos(30°))} = \boxed{1.8\frac{m}{s}}$$

The mass impacts the speed only in that if the mass is too small, air resistance and the mass of the string will become significant.

REFLECT

The tension in the string is always perpendicular to the ball's motion; thus, the work done by the tension will be zero.

Get Help: Picture It – Mechanical Energy

6.83

SET UP

A skier leaves the starting gate at an elevation of 4212 m with an initial speed of 4.00 m/s. The end of the ski slope is at an elevation of 4017 m. We are told that air resistance causes a 50% loss in the final kinetic energy (compared to the case with no air resistance). We can use conservation of mechanical energy to first determine the skier's final speed in the absence of air resistance (that is, the ideal case) and use this to calculate the skier's final speed taking air resistance into account (that is, the real case).

Figure 6-6 Problem 83

SOLVE

Conservation of mechanical energy:

$$U_i + K_i = U_f + K_f$$

$$\frac{1}{2}mv_i^2 + mgy_i = \frac{1}{2}mv_{f,\,ideal}^2 + mgy_f$$

$$v_{f,\,ideal}^2 = v_i^2 + 2g(y_i - y_f)$$

Considering air resistance:

$$K_{real} = \frac{1}{2}K_{ideal}$$

$$\frac{1}{2}mv_{f,\,real}^2 = \frac{1}{2}\left(\frac{1}{2}mv_{f,\,ideal}^2\right)$$

$$v_{f,\,real} = \sqrt{\frac{v_{f,\,ideal}^2}{2}} = \sqrt{\frac{v_i^2 + 2g(y_i - y_f)}{2}}$$

$$= \sqrt{\frac{\left(4.00\frac{m}{s}\right)^2 + 2\left(9.8\frac{m}{s^2}\right)((4212\text{ m}) - (4017\text{ m}))}{2}} = \boxed{43.8\frac{m}{s}}$$

REFLECT

The specifics regarding the middle of the skier's trip are not important since the surfaces are frictionless and we are told the change in energy caused by air resistance.

6.87

SET UP

The coefficient of restitution e of a ball is defined as the speed of the ball after impact divided by the speed of the ball just before impact. We can use conservation of mechanical energy to determine these speeds in terms of the heights h and H. The kinetic energy of the ball at the heights h and H is equal to zero since the ball is momentarily at rest.

SOLVE

Part a)
Before:

$$mgH = \frac{1}{2}mv^2_{\text{before}}$$

$$v_{\text{before}} = \sqrt{2gH}$$

After:

$$\frac{1}{2}mv^2_{\text{after}} = mgh$$

$$v_{\text{after}} = \sqrt{2gh}$$

Finding the coefficient of restitution:

$$e = \frac{v_{\text{after}}}{v_{\text{before}}} = \frac{\sqrt{2gh}}{\sqrt{2gH}} = \boxed{\sqrt{\frac{h}{H}}}$$

Part b)

$$e = \sqrt{\frac{h}{H}} = \sqrt{\frac{60 \text{ cm}}{80 \text{ cm}}} = \boxed{\sqrt{\frac{3}{4}} = 0.87}$$

REFLECT

A coefficient of restitution of 1 refers to a completely elastic collision.

6.91

SET UP

A 12-kg block is released from rest on a frictionless incline that makes an angle of 28 degrees with the horizontal. The block starts at some initial height h_1. The block slides down the ramp, comes into contact with a spring ($k = 13{,}500$ N/m), and comes to rest once the spring compresses a distance of 5.5 cm. Since there is no nonconservative work done on the block, we can use conservation of mechanical energy. The block starts and ends at rest, which means the kinetic energies are zero and the initial potential energy will equal the final potential energy. Initially, the block is not in contact with the spring, so the initial potential energy is all gravitational. If we define the height of the compressed spring to have a gravitational potential energy of zero, the final potential energy is equal to the potential energy stored in the spring. Setting these two equal, we can find the initial height of the block relative to its stopping point. Trigonometry will allow us to find the distance that the block traveled down the ramp.

Figure 6-7 Problem 91

SOLVE

$$U_i + K_i = U_f + K_f$$

$$mgh_i + 0 = \frac{1}{2}kx^2 + 0$$

$$h_i = \frac{kx^2}{2mg}$$

$$d = \frac{h_i}{\sin(28°)} = \frac{kx^2}{2mg\sin(28°)} = \frac{\left(13,500\frac{N}{m}\right)(0.055 \text{ m})^2}{2(12 \text{ kg})\left(9.8\frac{m}{s^2}\right)\sin(28°)} = \boxed{0.37 \text{ m}}$$

REFLECT

The block starts out a distance of 31.5 cm from the front of the uncompressed spring.

Get Help: Interactive Example – Block, Inclined Plane & Spring
P'Cast 6.13 – Marble on a String

6.95

SET UP

An object is released from rest on a frictionless ramp at a height of $H_1 = 12$ m. The bottom of this ramp merges smoothly with another frictionless ramp that makes an angle of $\theta_2 = 37$ degrees with the horizontal. Because both ramps are frictionless, the mechanical energy of the object is conserved. The block starts and ends at rest, which means the block will rise to the same height on the second ramp regardless of the angle of the ramp. We can use trigonometry to determine the distance the object travels up the second ramp. In part (b), the block starts from rest at a height of $H_1 = 12$ m; we can use conservation of mechanical energy to calculate its speed at a height of $H_2 = 7$ m.

Figure 6-8 Problem 95

SOLVE

Part a)

$$d_2 = \frac{H_2}{\sin(37°)} = \frac{12 \text{ m}}{\sin(37°)} = \boxed{20 \text{ m}}$$

Part b)

$$U_i + K_i = U_f + K_f$$

$$mgh_i + 0 = mgh_f + \frac{1}{2}mv_f^2$$

$$v_f = \sqrt{2g(h_i - h_f)} = \sqrt{2\left(9.8\frac{m}{s^2}\right)((12 \text{ m}) - (7 \text{ m}))} = \boxed{9.9\frac{m}{s}}$$

REFLECT

The angle of the first ramp is not used because the ramp is frictionless and mechanical energy is conserved.

Get Help: Picture It – Mechanical Energy

Chapter 7
Linear Momentum

Conceptual Questions

7.3 Yes, if the tennis ball moves 18 times faster.

Get Help: P'Cast 7.1 – Momentum of Two Objects

7.7 Pushing off the boat to step onto the pier will push the boat away. You will make less progress than on solid ground and can easily fall in the gap.

7.15 (a) Choose three orthogonal directions, ideally so that at least one is aligned with a direction around which the sheet is symmetric. For any axis around which the sheet is symmetric, the center of mass in the perpendicular direction is along that axis. Otherwise, the center of mass in a given direction is found by adding up (that is, integrating) the product of the mass of infinitesimally small elements of the sheet and the distance each is from the axis and then dividing by the mass of the sheet. (b) The same process holds for a sheet with a hole, with the additional step that the hole is included in the process as a "negative" mass.

Get Help: Picture It – Center of Mass

Multiple-Choice Questions

7.19 D ($2mv$). If the initial momentum of the ball points in the $-x$ direction, its initial momentum is $-mv$ and the final momentum is $+mv$. This is a total change in momentum of $2mv$.

7.23 B (The magnitudes of the velocities are the same but the directions are reversed). By rearranging the energy and momentum conservation equations for two objects undergoing a one-dimensional, elastic collision, we can write the final speeds of each object as:

$$v_{1,\,f} = \left(\frac{m_1 - m_2}{m_1 + m_2}\right)v_{1,\,i} + \left(\frac{2m_2}{m_1 + m_2}\right)v_{2,\,i}$$

$$v_{2,\,f} = \left(\frac{2m_1}{m_1 + m_2}\right)v_{1,\,i} + \left(\frac{m_2 - m_1}{m_1 + m_2}\right)v_{2,\,i}$$

For the case of equal masses (that is, $m_1 = m_2$), we find that $v_{1,\,f} = v_{2,\,i}$ and $v_{2,\,f} = v_{1,\,i}$.

Get Help: Picture It – Momentum Conservation
P'Cast 7.8 – A Fly and an Elephant

7.27 A (Block A rises to a height greater than H_A and block B rises to a height less than H_B). Conservation of momentum:

$$m_A\sqrt{2gH_A} - m_B\sqrt{2gH_B} = m_A v_{A,\,f} + m_B v_{B,\,f}$$

Conservation of energy:

$$m_A(2gH_A) + m_B(2gH_B) = m_A v_{A, f}^2 + m_B v_{B, f}^2$$

Solving this system of equations for the final speeds and then applying conservation of energy to calculate the final heights, we'll find that the less massive block (block A) will rise to a greater height than the more massive block (block B).

We can also look at the extremes of the system. What if block B is way more massive than block A? If they are released from about the same heights, then certainly block A rises higher after the collision. If they are released so that block B is nearly at the bottom initially, then it's going slowly when they collide, perhaps nearly zero, but that, too, makes block A rise higher. What if block B is just slightly more massive than block A? If they are released from about the same heights, then they have about the same speeds when they collide, but that would result in block A having just a slightly higher speed afterward. So again, block A rises higher after the collision. And, finally, if block B is just slightly more massive than block A and block B is released from way down low, so that block B is going almost zero when they collide, since A is less massive than block B it rebounds at a slightly higher speed than its initial speed, which means it will rise higher.

Estimation Questions

7.31 Roughly speaking, it is around your navel. The ratio of the center of mass to height in humans is about 0.55.

Get Help: Picture It – Center of Mass

7.35 A tennis ball has a mass of 0.057 kg and a coefficient of restitution of 0.94. If an incoming volley is 40 m/s, then we get an impulse of around 5 kg · m/s.

Get Help: P'Cast 7.7 – Follow Through

Problems

7.39

SET UP
The magnitude of the instantaneous momentum of a 0.057-kg tennis ball is 2.6 kg · m/s. We can use the definition of momentum to calculate the instantaneous speed of the ball.

SOLVE

$$p = mv$$

$$v = \frac{p}{m} = \frac{\left(2.6\dfrac{\text{kg} \cdot \text{m}}{\text{s}}\right)}{0.057 \text{ kg}} = \boxed{46\dfrac{\text{m}}{\text{s}}}$$

REFLECT
We would expect a large speed since the mass of the ball is small. This is a reasonable speed for a tennis ball.

Get Help: Momentum of Two Objects

7.43

SET UP

A 55-kg girl is riding a skateboard at a speed of 6 m/s. We can calculate the magnitude of her momentum directly from the definition of momentum. We are given the magnitude of the skateboard's momentum and we know that it must be traveling at the same speed as the girl. Therefore, we can find the mass of the skateboard through division.

SOLVE

Part a)

$$p = mv = (55 \text{ kg})\left(6\frac{\text{m}}{\text{s}}\right) = \boxed{330\frac{\text{kg} \cdot \text{m}}{\text{s}}}$$

Part b)

$$m = \frac{p}{v} = \frac{\left(30\dfrac{\text{kg} \cdot \text{m}}{\text{s}}\right)}{\left(6\dfrac{\text{m}}{\text{s}}\right)} = \boxed{5 \text{ kg}}$$

REFLECT

The momentum of the skateboard is one-eleventh of the momentum of the girl, which means the mass of the skateboard will be one-eleventh of the mass of the girl.

7.47

Figure 7-1 Problem 47

SET UP

An object of mass $3M$ has an initial velocity of $\vec{v}_0 = v_0\hat{x}$. It breaks into two unequal pieces. Piece 1 has a mass of M and travels off at an angle of 45 degrees below the x-axis with a speed of v_1. Piece 2 has a mass of $2M$ and travels off at an angle of 30 degrees above the x-axis with a speed of v_2. We can use conservation of momentum in order to calculate the final velocities of the two pieces. Since this is a two-dimensional problem, we will need to split the momenta into components and solve the x and y component equations.

SOLVE

y component:

$$0 = (2M)v_2\sin(30°) - (M)v_1\sin(45°)$$

$$v_2 = \frac{v_1\sin(45°)}{2\sin(30°)} = \frac{v_1}{\sqrt{2}}$$

x component:

$$(3M)v_0 = (2M)v_2\cos(30°) + (M)v_1\cos(45°)$$

$$3v_0 = 2v_2\cos(30°) + v_1\cos(45°) = 2\left(\frac{v_1}{\sqrt{2}}\right)\left(\frac{\sqrt{3}}{2}\right) + \frac{v_1}{\sqrt{2}}$$

$$v_1 = \frac{3\sqrt{2}}{\sqrt{3}+1}v_0 = 1.55v_0$$

$$v_2 = \frac{v_1}{\sqrt{2}} = \frac{3}{\sqrt{3}+1}v_0 = 1.10v_0$$

Therefore,

$$\vec{v}_1 = (1.55v_0\cos(45°))\hat{x} - (1.55v_0\sin(45°))\hat{y} = \boxed{(1.1v_0)\hat{x} - (1.1v_0)\hat{y}}$$

$$\vec{v}_2 = (1.10v_0\cos(30°))\hat{x} + (1.10v_0\sin(30°))\hat{y} = \boxed{(0.95v_0)\hat{x} + (0.55v_0)\hat{y}}$$

REFLECT

Initially there is no momentum in the *y* direction, which means the final *y* momentum must also be zero. If we calculate it explicitly, we see that this is true: $(M)(-1.1v_0) + (2M)(0.55v_0) = 0$.

Get Help: P'Cast 7.5 – Superheroes Conserve Momentum

7.51

SET UP

Two bighorn sheep are in the midst of a head-butting contest. Sheep 1 ($m_1 = 95$ kg) moves at a speed of 10 m/s directly toward sheep 2 ($m_2 = 80$ kg) running at 12 m/s. To determine which sheep wins the contest, we need to compare the momenta of the two sheep; whichever sheep has the larger momentum will end up knocking the other sheep backward.

SOLVE

Sheep 1:

$$p_1 = m_1 v_1 = (95 \text{ kg})\left(10\frac{\text{m}}{\text{s}}\right) = 950\frac{\text{kg} \cdot \text{m}}{\text{s}}$$

Sheep 2:

$$p_2 = m_2 v_2 = (80 \text{ kg})\left(12\frac{\text{m}}{\text{s}}\right) = 960\frac{\text{kg} \cdot \text{m}}{\text{s}}$$

The lighter sheep has a larger momentum, so $\boxed{\text{sheep 2}}$ will win the head-butting contest.

REFLECT

If we knew how long the sheep were in contact, we could calculate the force of one sheep on the other. We could then relate this to the acceleration of the sheep via Newton's second law. If the net force on the sheep were constant, we could use the constant acceleration equations to calculate how far sheep 2 pushes sheep 1.

7.55

SET UP

A howler monkey ($m_1 = 5.0$ kg) is swinging from a vine at a speed of $v_{1,i} = 12$ m/s due east toward a second monkey ($m_2 = 6.0$ kg) that is also moving east at a speed of $v_{2,i} = 8$ m/s. The monkeys grab onto one another and travel together on the same vine. We can calculate the final speed v_f of the two monkeys traveling together by applying conservation of momentum. In our coordinate system, east will point in the positive direction.

SOLVE

$$m_1 v_{1,i} + m_2 v_{2,i} = (m_1 + m_2)v_f$$

$$v_f = \frac{m_1 v_{1,i} + m_2 v_{2,i}}{m_1 + m_2} = \frac{(5.0 \text{ kg})\left(12\frac{m}{s}\right) + (6.0 \text{ kg})\left(8\frac{m}{s}\right)}{(5.0 \text{ kg}) + (6.0 \text{ kg})} = \boxed{9.8\frac{m}{s}}$$

REFLECT

It makes sense that the monkeys will continue to travel east at a speed in between their initial speeds.

Get Help: P'Cast 7.6 – Car Crash

7.59

SET UP

A 0.200-kg ball is initially traveling at 20 m/s toward your hand. You catch the ball and it comes to rest within 0.025 s. We can find the magnitude of the force your hand exerts on the ball from the change in the ball's momentum and the contact time.

SOLVE

$$F = \left|\frac{\Delta p}{\Delta t}\right| = \left|\frac{m(v_f - v_i)}{\Delta t}\right| = \left|\frac{(0.200 \text{ kg})\left(0 - \left(20\frac{m}{s}\right)\right)}{0.025 \text{ s}}\right| = \boxed{160 \text{ N}}$$

REFLECT

The force of your hand on the ball will act opposite to the ball's initial velocity.

Get Help: P'Cast 7.7 – Follow Through

7.63

SET UP

We are given the functional form for a one-dimensional force as a function of time $F(t)$ in SI units. This force acts on an object that is initially at rest, which means its initial momentum is zero. We can calculate the momentum of the object after the force has acted for 1 s by integrating $F(t)$ with respect to time from $t = 0$ s to $t = 1$ s.

SOLVE

$$\Delta p = p_f - p_i = p_f - 0 = \int_0^1 F(t)dt$$

$$p_f = \int_0^1 F(t)dt = \int_0^1 (9t^2 + 8t^3)dt = [3t^3 + 2t^4]_0^1 = [3(1)^3 + 2(1)^4 - 0 - 0] = \boxed{5\frac{kg \cdot m}{s}}$$

REFLECT

The momentum of the object points in the same direction as the force

7.67

SET UP

A block of ice ($m_1 = 10$ kg) is moving at 8 m/s toward the east when it collides elastically with a second block of ice ($m_2 = 6$ kg) moving toward the east at 4 m/s. We can use conservation of momentum and conservation of mechanical energy to find the final velocities of the two blocks. We can derive a useful relationship between the final velocities and the initial velocities of a one-dimensional elastic collision: $v_{1,f} = \left(\dfrac{m_1 - m_2}{m_1 + m_2}\right)v_{1,i} + \left(\dfrac{2m_2}{m_1 + m_2}\right)v_{2,i}$ and $v_{2,f} = \left(\dfrac{2m_1}{m_1 + m_2}\right)v_{1,i} + \left(\dfrac{m_2 - m_1}{m_1 + m_2}\right)v_{2,i}$. In our coordinate system, we'll consider $+x$ to point toward the east.

SOLVE

$$v_{1,f} = \left(\frac{m_1 - m_2}{m_1 + m_2}\right)v_{1,i} + \left(\frac{2m_2}{m_1 + m_2}\right)v_{2,i}$$

$$= \left(\frac{(10\text{ kg}) - (6\text{ kg})}{(10\text{ kg}) + (6\text{ kg})}\right)\left(8\frac{m}{s}\right) + \left(\frac{2(6\text{ kg})}{(10\text{ kg}) + (6\text{ kg})}\right)\left(4\frac{m}{s}\right) = 5\frac{m}{s}$$

Block 1's final velocity is $\boxed{5 \text{ m/s to the right}}$.

$$v_{2,f} = \left(\frac{2m_1}{m_1 + m_2}\right)v_{1,i} + \left(\frac{m_2 - m_1}{m_1 + m_2}\right)v_{2,i}$$

$$= \left(\frac{2(10\text{ kg})}{(10\text{ kg}) + (6\text{ kg})}\right)\left(8\frac{m}{s}\right) + \left(\frac{(6\text{ kg}) - (10\text{ kg})}{(10\text{ kg}) + (6\text{ kg})}\right)\left(4\frac{m}{s}\right) = 9\frac{m}{s}$$

Block 2's final velocity is $\boxed{9 \text{ m/s to the right}}$.

REFLECT

It makes sense that block 1 should slow down and block 2 should move faster after the collision.

Get Help: Picture It – Momentum Conservation
P'Cast 7.8 – A Fly and an Elephant

7.71

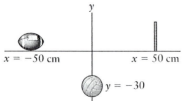

Figure 7-2 Problem 71

SET UP

The locations of three uniform objects (a rod, a football, and a volleyball) are given. Since the objects are uniform, the center of mass of each object is located at its geometrical center. The center of an ellipse is halfway along each axis. (Remember that the semimajor axis is the longer axis.)

SOLVE

Rod:

$$\boxed{\begin{aligned} x_{CM} &= 50 \text{ cm} \\ y_{CM} &= 15 \text{ cm} \end{aligned}}$$

Football:

$$x_{CM} = -50 \text{ cm}$$
$$y_{CM} = 4 \text{ cm}$$

Volleyball:

$$x_{CM} = 0 \text{ cm}$$
$$y_{CM} = -30 \text{ cm}$$

REFLECT

Because we knew the dimensions of each shape and that they were uniform, we did not need to know the mass of the objects. We would get the same answers if we explicitly calculated the center of mass in each case.

Get Help: Picture It – Center of Mass

7.75

SET UP

Three hundred million people fall from a height of 1 m onto the ground. We can use conservation of mechanical energy to calculate the speed with which the people land on Earth. From this speed, we can calculate the magnitude of the momentum of all 300 million people; this is the momentum that will be imparted to Earth. Dividing this quantity by the mass of Earth will give us the change in speed of Earth.

SOLVE

Part a)
Speed of a single person:

$$mgh = \frac{1}{2}mv^2$$

$$v = \sqrt{2gh} = \sqrt{2\left(9.8\frac{m}{s^2}\right)(1 \text{ m})} = 4.4\frac{m}{s}$$

Momentum of 300 million people:

$$p = mv = (3.00 \times 10^8)(65 \text{ kg})\left(4.4\frac{m}{s}\right) = \boxed{8.6 \times 10^{10}\frac{\text{kg} \cdot \text{m}}{s}}$$

Part b)

$$\Delta p_E = m_E \Delta v_E$$

$$\Delta v_E = \frac{\Delta p_E}{m_E} = \frac{\left(8.6 \times 10^{10}\frac{\text{kg} \cdot \text{m}}{s}\right)}{6.0 \times 10^{24} \text{ kg}} = \boxed{1.445 \times 10^{-14}\frac{m}{s}}$$

REFLECT

We could have also used kinematics to calculate the landing speed of the people. Technically Earth will recoil at a nonzero speed, but a speed on the order of 10^{-14} m/s is imperceptible, so Earth effectively stays still.

7.79

SET UP

A train car ($m_t = 8000$ kg) is rolling at an initial speed of $v_i = 20.0$ m/s when, all of a sudden, rainwater collects inside of it. The final speed of the filled train car is $v_f = 19$ m/s. We can calculate the mass of the rainwater from conservation of momentum.

SOLVE

$$m_t v_i = (m_t + m_w)v_f$$

$$m_w = \frac{m_t v_i}{v_f} - m_t = \frac{(8000 \text{ kg})\left(20.0\dfrac{\text{m}}{\text{s}}\right)}{\left(19\dfrac{\text{m}}{\text{s}}\right)} - (8000 \text{ kg}) = \boxed{421 \text{ kg}}$$

REFLECT

The speed of the train car decreased by about 5%, which means the mass must have increased by about 5%.

Get Help: P'Cast 7.6 – Car Crash

7.83

SET UP

A lion ($m_L = 135$ kg) is running northward at 80 km/hr when it collides and latches onto a gazelle ($m_G = 29$ kg) that is running eastward at 60 km/hr. In our coordinate system east will point toward $+x$ and north will point toward $+y$. We can use conservation of momentum to find the final speed and direction of the lion–gazelle system after the lion attacks. Since this is a two-dimensional problem, we will need to split the momenta into components and solve the x and y component equations.

SOLVE

x component:

$$m_L v_{L,x} + m_G v_{G,x} = (m_L + m_G)v_{f,x}$$

$$v_{f,x} = \frac{m_L v_{L,x} + m_G v_{G,x}}{(m_L + m_G)} = \frac{0 + (29 \text{ kg})\left(60\dfrac{\text{km}}{\text{hr}}\right)}{(135 \text{ kg}) + (29 \text{ kg})} = 11\frac{\text{km}}{\text{hr}}$$

y component:

$$m_L v_{L,y} + m_G v_{G,y} = (m_L + m_G)v_{f,y}$$

$$v_{f,y} = \frac{m_L v_{L,y} + m_G v_{G,y}}{(m_L + m_G)} = \frac{(135 \text{ kg})\left(80\dfrac{\text{km}}{\text{hr}}\right) + 0}{(135 \text{ kg}) + (29 \text{ kg})} = 66\frac{\text{km}}{\text{hr}}$$

Final speed:

$$v = \sqrt{v_{f,x}^2 + v_{f,y}^2} = \sqrt{\left(11\frac{\text{km}}{\text{hr}}\right)^2 + \left(66\frac{\text{km}}{\text{hr}}\right)^2} = \boxed{67\frac{\text{km}}{\text{hr}}}$$

Final direction:

$$\theta = \arctan\left(\frac{66}{11}\right) = \boxed{80.5° \text{ north of east}}$$

REFLECT

The momentum of the lion is considerably larger than the momentum of the gazelle, so the final momentum of the system should point more north than east.

Get Help: P'Cast 7.6 – Car Crash

7.85

Figure 7-3 Problem 85

SET UP

A bullet ($m_B = 0.012$ kg) is shot into a block of wood of mass m_W at a speed of $v_{B\,i} = 250$ m/s. The wood is attached to a spring with a spring constant $k = 200$ N/m. Once the bullet embeds itself into the wood, the spring compresses a distance of $x = 30$ cm before coming to a stop. We first need to use conservation of momentum in order to calculate the speed of the (bullet + wood) after the collision in terms of m_W. Then we can use conservation of mechanical energy to solve for the numerical value of m_W.

SOLVE

Conservation of momentum:

$$m_B v_{B,\,i} + m_W v_{W,\,i} = (m_B + m_W)v_f$$

$$v_f = \frac{m_B v_{B,\,i} + m_W v_{W,\,i}}{m_B + m_W} = \frac{m_B v_{B,\,i} + 0}{m_B + m_W} = \frac{m_B v_{B,\,i}}{m_B + m_W}$$

Conservation of mechanical energy:

$$U_i + K_i = U_f + K_f$$

$$0 + \frac{1}{2}(m_B + m_W)v_f^2 = \frac{1}{2}kx^2 + 0$$

$$(m_B + m_W)v_f^2 = kx^2$$

$$(m_B + m_W)\left(\frac{m_B v_{B,\,i}}{m_B + m_W}\right)^2 = kx^2$$

$$\frac{1}{(m_B + m_W)}(m_B v_{B,\,i})^2 = kx^2$$

$$m_W = \frac{(m_B v_{B,\,i})^2}{kx^2} - m_B = \frac{(0.012\ \text{kg})^2\left(250\dfrac{\text{m}}{\text{s}}\right)^2}{\left(200\dfrac{\text{N}}{\text{m}}\right)(0.30\ \text{m})^2} - (0.012\ \text{kg}) = \boxed{0.49\ \text{kg}}$$

REFLECT

The block of wood with the embedded bullet moves at a speed of 5.9 m/s.

7.93

SET UP

A uniform sheet of metal of total mass M is formed into an isosceles triangle that is 1 m long in the x direction and 2 m long in the y direction. Since this is a solid, continuous object, we need to use calculus to find the center of mass coordinates of the triangle: $x_{CM} = \dfrac{\int x\,dm}{M}$ and

$y_{CM} = \dfrac{\int y\,dm}{M}$. We split the triangle up into small rectangles of area dA. We can convert the

integral from dm to dA through the density.

Height = 2 m
Base = 1 m

Figure 7-4 Problem 93

SOLVE

x component:

The triangle is symmetric about the y-axis; therefore,

$$\boxed{x_{CM} = 0}$$

y component:

$$\frac{M}{A} = \frac{dm}{dA}$$

$$dm = \frac{M}{A}dA = \frac{M}{\left(\dfrac{1}{2}\right)(1)(2)}(2x\,dy) = 2Mx\,dy$$

But, through similar triangles,

$$\frac{2-y}{x} = \frac{2}{0.5}$$

Therefore,

$$dm = 2Mx\,dy = 2M\left(\frac{2-y}{4}\right)dy = \frac{M}{2}(2-y)dy$$

$$y_{CM} = \frac{1}{M}\int y\,dm = \frac{1}{M}\int_0^2 My\frac{(2-y)}{2}dy = \frac{1}{2}\int_0^2 (2y - y^2)dy = \frac{1}{2}\left[y^2 - \frac{y^3}{3}\right]_0^2$$

$$= \frac{1}{2}\left[(2)^2 - \frac{(2)^3}{3}\right] = \left[2 - \frac{4}{3}\right] = \boxed{0.67 \text{ m}}$$

REFLECT

More of the mass is located near the bottom of the triangle, so we would expect y_{CM} to be less than half the height.

Get Help: Picture It – Center of Mass

Chapter 8
Rotational Motion

Conceptual Questions

8.3 The moment of inertia of a rotating object is calculated relative to the axis of rotation. If we don't explicitly define this axis at the outset of the problem, the calculated moment of inertia will be ambiguous at best and meaningless at worst. Even for axes crossing the center of mass, different axes will have different moments.

Get Help: Picture It – Moment of Inertia

8.7 When riding a seesaw, the torque produced by each rider about the pivot should be approximately equal in magnitude. The riders are trying to move their centers of mass relative to the pivot of the seesaw by leaning forward or backward. This, in turn, will affect the magnitude of the torque about the pivot produced by each rider by changing the moment arms. We are told that the riders are sitting equidistant from the pivot point and the person at the top always leans backward, which means she has a smaller mass than the person at the bottom (who always leans forward).

Get Help: Interactive Example – Pivot Rod with Weight

8.11 The term *dm* represents a tiny piece of mass from a large-scale object.

8.15 If we consider the system to be the cookie dough + turntable, there is **no net torque** acting on the system, and the **rotational kinetic energy remains constant** since there is no net work done on the system. This means the **angular momentum of the system is constant.** Dropping the cookie dough onto the edge of the turntable changes the way the mass in the system is distributed about the axis of rotation, thus **changing the moment of inertia.** Since the angular momentum is constant and the moment of inertia of the system changes, the **angular velocity of the system must change** after the cookie dough lands. (The moment of inertia increases. Therefore, the **magnitude of the angular velocity decreases**; that is, the turntable slows down.) A changing angular velocity means there is a **nonzero angular acceleration.**

Get Help: Picture It – Angular Momentum
Interactive Example – Pivot Putty

8.19 We can use conservation of energy to answer this question. All three objects—a sphere, a cylinder, and a ring—start at rest from the same height and, thus, have the same initial potential energy. As the objects roll down the inclined plane, potential energy is converted into rotational *and* translational kinetic energy. The smaller the moment of inertia of an object, the more energy that will be converted into translational kinetic energy rather than rotational kinetic energy. The object that reaches the bottom of the ramp has the largest translational kinetic energy, so the sphere $\left(I_{\text{sphere}} = \frac{2}{5} MR^2 \right)$ "wins," followed by the cylinder $\left(I_{\text{cylinder}} = \frac{1}{2} MR^2 \right)$, and then the hoop ($I_{\text{hoop}} = MR^2$).

8.25 Counterclockwise corresponds to the positive sense of rotation. We're told the merry-go-round undergoes constant acceleration.

a) Graph (c). The angular speed increases linearly as a function of time.

b) Graph (i). The angular displacement increases parabolically under constant angular acceleration.

c) Graph (a). The merry-go-round maintains a constant maximum angular speed.

d) Graph (ii). The angular displacement increases linearly for a constant angular speed.

e) Graph (d). The angular speed should decrease linearly from a maximum to zero.

f) Graph (iv). The angular position of the merry-go-round will eventually level out once it comes to a stop.

g) When the merry-go-round is speeding up in the counterclockwise direction, there is a positive angular acceleration caused by a positive net torque on the merry-go-round. Since we're told that the angular acceleration is constant, the net torque must also be constant. The net torque is equal to zero when the merry-go-round is spinning at a constant angular velocity. The merry-go-round slows down due to a constant, negative net torque being applied.

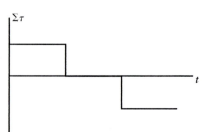

Figure 8-1 Problem 25

Multiple-Choice Questions

8.31 **C** (the hoop). Three objects—a solid ball, a solid disk, and a hoop—that have the same mass and radius start rolling *and* translating up an incline from the same height, which we'll call $y = 0$. The objects all begin moving at the same linear speed. The moment of inertia for each of the objects is different, though:

$$I_{\text{solid sphere}} = \frac{2}{5}MR^2, \; I_{\text{solid disk}} = \frac{1}{2}MR^2, \; I_{\text{hoop}} = MR^2.$$

We can use conservation of mechanical energy to solve this problem:

$$K_{\text{rot, i}} + K_{\text{tr, i}} + U_{\text{g, i}} = K_{\text{rot, f}} + K_{\text{tr, f}} + U_{\text{g, f}}$$

$$\frac{1}{2}I\omega^2 + \frac{1}{2}mv^2 + 0 = 0 + 0 + mgh$$

$$\frac{1}{2}I\left(\frac{v}{R}\right)^2 + \frac{1}{2}mv^2 = mgh$$

Solving for the height h that each object makes it up the incline, $h = \dfrac{v^2}{2mg}\left(\dfrac{I}{R^2} + m\right)$.

Since the moment of inertia for the hoop has the largest coefficient (1 versus $\frac{2}{5}$ versus $\frac{1}{2}$), it will go up the incline the farthest.

8.35 **B** ($2MR^2$). We need to find the moment of inertia of a thin ring of radius R rotating about a point on its rim. The parallel-axis theorem will give us the moment of inertia of an object rotating about an axis *not* passing through its center of mass: $I = I_{\text{CM}} + Mb^2$, where I_{CM} is the moment

Rotation axis

R

Figure 8-2 Problem 35

of inertia of the object rotating about an axis through its center of mass and h is the distance between that axis and the actual rotation axis. For a thin ring, $I_{CM} = MR^2$. The distance from the center of mass to a point on the edge of the ring is R. Therefore, $I = I_{CM} + Mh^2 = MR^2 + MR^2 = 2MR^2$.

Estimation Questions

8.37 We're told that the cloverleaf is approximately three-quarters of a circle, which corresponds to a total angular displacement of $\Delta\theta = \frac{3}{4}(2\pi \text{ rad}) = \frac{3\pi}{2}\text{rad}$. We need to estimate the amount of time the car is driving on the cloverleaf. Let's say that the cloverleaf is three-quarters of a circle of 25 m in radius, and the car is driving a little over 30 mph (15 m/s). The total

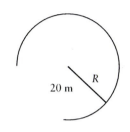

Figure 8-3 Problem 37

distance the car travels is the arc length, or $\left(\frac{3\pi}{2}\text{rad}\right)(25 \text{ m}) = \frac{75\pi}{2}\text{m}$. It takes the

car $\dfrac{\left(\dfrac{75\pi}{2}\text{m}\right)}{\left(15\dfrac{\text{m}}{\text{s}}\right)} = \dfrac{5\pi}{2}\text{s}$ to drive around the cloverleaf. The angular speed is the angular

displacement over the time it took, so $\omega = \dfrac{\Delta\theta}{\Delta t} = \dfrac{\left(\dfrac{3\pi}{2}\text{rad}\right)}{\left(\dfrac{5\pi}{2}\text{s}\right)} = \boxed{\dfrac{3}{5}\dfrac{\text{rad}}{\text{s}} = 0.6\dfrac{\text{rad}}{\text{s}}}$.

8.45 The magnitude of the "skip-it" ball's angular momentum is equal to $L = I\omega = MR^2\omega$, where we plugged in the moment of inertia for a point mass. Since the toy is made of plastic and is attached to a child's leg, the mass of the ball should be around 0.25 kg. The toy is approximately a meter long. The ball makes about a revolution per second when the child is swinging it around her leg; this corresponds to an angular speed of $\omega = 1\dfrac{\text{rev}}{\text{s}} \times \dfrac{2\pi \text{ rad}}{1 \text{ rev}} = 2\pi\dfrac{\text{rad}}{\text{s}}$. Plugging all of this in yields $L = (0.25 \text{ kg})(1 \text{ m})^2 \times$ $\left(2\pi\dfrac{\text{rad}}{\text{s}}\right) = \dfrac{\pi}{2}\dfrac{\text{kg} \cdot \text{m}^2}{\text{s}} \approx \boxed{1.6\dfrac{\text{kg} \cdot \text{m}^2}{\text{s}}}$.

Get Help: Interactive Example – Pivot Putty
P'Cast 8.5 – Figure Skater

Problems

8.51

SET UP

It is reasonable to assume that the angular speed of the Moon is roughly constant. In this case, the angular speed is given by the angular displacement divided by the time it took. We're told that the Moon completes one orbit (assuming a circular orbit) in 27.4 days. One circular orbit is an angular displacement of 2π radians.

SOLVE

$$1 \text{ orbit} = 27.4 \text{ day} \times \frac{24 \text{ hr}}{1 \text{ day}} \times \frac{3600 \text{ s}}{1 \text{ hr}} = 2.367 \times 10^6 \text{ s}$$

$$\omega = \frac{\Delta\theta}{\Delta t} = \frac{2\pi \text{ rad}}{2.367 \times 10^6 \text{ s}} = \boxed{2.65 \times 10^{-6} \frac{\text{rad}}{\text{s}}}$$

REFLECT

Since it takes such a long time for the Moon to complete one orbit, we expect its angular speed to be a small number. The Earth–Moon distance is not necessary for finding the angular speed.

8.55

SET UP

We are given the moment of inertia I and rotational kinetic energy K_{rot} of an object and asked to find its angular speed. We can use the definition of rotational kinetic energy, $K_{rot} = \frac{1}{2}I\omega^2$, and solve for ω. To convert from radians per second, we need to realize that one revolution corresponds to an angular displacement of 2π radians.

SOLVE

Solving for ω and plugging in the given values, we find that

$$\omega = \sqrt{\frac{2K_{rot}}{I}} = \sqrt{\frac{2(2.75 \text{ J})}{0.33 \text{ kg} \cdot \text{m}^2}}$$

$$= 4.08\frac{\text{rad}}{\text{s}} \times \frac{60 \text{ s}}{1 \text{ min}} \times \frac{1 \text{ rev}}{2\pi \text{ rad}} = \boxed{39.0\frac{\text{rev}}{\text{min}}}$$

REFLECT

Since we are given the moment of inertia explicitly, we do not need to use the fact that it is a rotating wheel.

8.61

SET UP

The steering wheel is rotating about an axis that passes through its center and is made up of a rim of radius R and mass M and five thin rods of length R and mass $M/2$. The overall moment of inertia of the steering wheel is equal to the moment of inertia of the rim plus the moments of inertia for each of the thin rods. We can model the rim as a uniform ring of radius R rotating about an axis through its center, and each of the thin rods is rotating about a perpendicular line through one end.

Figure 8-4 Problem 61

SOLVE

$$I_{rim} = I_{ring} = MR^2$$

$$I_{rod} = \frac{1}{3}\left(\frac{M}{2}\right)R^2 = \frac{1}{6}MR^2$$

$$I_{total} = I_{rim} + 5I_{rod} = MR^2 + 5\left(\frac{1}{6}MR^2\right) = MR^2 + \frac{5}{6}MR^2 = \boxed{\frac{11}{6}MR^2}$$

REFLECT

Since the steering wheel is made up of simple geometric objects, we can use the results in Table 8-1 of the textbook. We then determine the axis of rotation for the overall object. This allows us to determine which moments of inertia to use (for example, a thin rod rotating about an axis through its center versus a thin rod rotating about an axis through one end).

8.67

SET UP

The system is made up of three separate objects: a solid sphere of radius R and mass M, another solid sphere of radius $2R$ and mass M, and a thin, uniform rod of length $3R$ and mass M. We need to find the moment of inertia about the axis through the center of the rod by considering the moments of inertia for each of the component objects in order to find the overall moment of inertia of the system. Since the two spheres are not located at the axis of rotation, we need to determine the distance their centers of mass are from the axis of rotation and invoke the parallel-axis theorem. The center of mass for the large sphere is $3.5R$ from the axis of rotation, and the small sphere is $2.5R$ from the axis.

Figure 8-5 Problem 67

SOLVE

$$I_{total} = I_{rod} + I_{left\ sphere} + I_{right\ sphere}$$

$$I_{rod} = \frac{1}{12}M(3R)^2 = \frac{9}{12}MR^2$$

$$I_{left\ sphere} = I_{sphere} + I_{from\ axis} = \frac{2}{5}M(2R)^2 + M\left(2R + \frac{3}{2}R\right)^2 = \frac{8}{5}MR^2 + \frac{49}{4}MR^2 = \frac{277}{20}MR^2$$

$$I_{right\ sphere} = I_{sphere} + I_{from\ axis} = \frac{2}{5}MR^2 + M\left(R + \frac{3}{2}R\right)^2 = \frac{2}{5}MR^2 + \frac{25}{4}MR^2 = \frac{133}{20}MR^2$$

$$I_{total} = I_{rod} + I_{left\ sphere} + I_{right\ sphere} = \frac{9}{12}MR^2 + \frac{277}{20}MR^2 + \frac{133}{20}MR^2 = \boxed{\frac{255}{12}MR^2}$$

REFLECT

The system is made up of simple geometric objects, so we can use the results in Table 8-1 of the textbook. We need to be careful to add the distance the center of mass of each sphere is from the rotation axis in determining the overall moment of inertia. Since the solid spheres are not located along the axis of rotation, it makes sense that the overall moment of inertia of the system is larger than just the sum of

$$I_{rod} + I_{sphere\ of\ 2R} + I_{sphere\ of\ R} = \frac{9}{12}MR^2 + \frac{8}{5}MR^2 + \frac{2}{5}MR^2 = \frac{33}{12}MR^2$$

8.71

SET UP

A pencil of length L is made up of a piece of wood of mass M and length $7L/8$ and an eraser of mass M and length $L/8$. We need to determine the moment of inertia of the pencil about an

axis through the end of the eraser using integration: $I = \int r^2 dm$. The overall moment of inertia of the pencil is the moment of inertia due to the eraser plus the moment of inertia due to the wood. We can model the pencil as a uniform, thin rod with the end of the eraser located at $y = 0$ and the top of the wooden part at $y = L$ to make the calculation easier because we only need to consider one dimension. In this coordinate system the eraser starts at $y = 0$ and ends at $y = L/8$ and the wood starts at $y = L/8$ and ends at $y = L$. Since we are assuming the pencil is uniform, the density throughout each part (eraser, wood) is constant. This allows us to convert our integration from dm to dy.

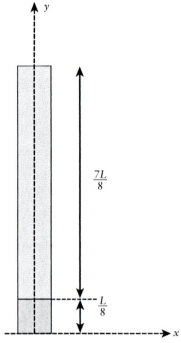

Figure 8-6 Problem 71

SOLVE

$$I_{pencil} = \int_0^L y^2 dm = \int_0^{L/8} y^2 dm_{eraser} + \int_{L/8}^L y m_{wood}$$

We can find dm_{eraser} and dm_{wood} by remembering that

$$\frac{\text{length of slice}}{\text{length of eraser}} = \frac{\text{mass of slice}}{\text{mass of eraser}} \quad \text{and} \quad \frac{\text{length of slice}}{\text{length of wood}} = \frac{\text{mass of slice}}{\text{mass of wood}}.$$

Therefore,

$$dm_{eraser} = \frac{M}{\left(\dfrac{L}{8}\right)} dy = \frac{8M}{L} dy \quad \text{and} \quad dm_{wood} = \frac{M}{\left(\dfrac{7L}{8}\right)} dy = \frac{8M}{7L} dy$$

Plugging these into the integral,

$$I_{pencil} = \int_0^{L/8} y^2 dm_{eraser} + \int_{L/8}^L y^2 dm_{wood} = \int_0^{L/8} y^2\left(\frac{8M}{L}\right) dy + \int_{L/8}^L y^2\left(\frac{8M}{7L}\right) dy$$

$$= \left(\frac{8M}{L}\right)\int_0^{L/8} y^2 dy + \left(\frac{8M}{7L}\right)\int_{L/8}^L y^2 dy = \left(\frac{8M}{L}\right)\left[\frac{y^3}{3}\right]_0^{L/8} + \left(\frac{8M}{7L}\right)\left[\frac{y^3}{3}\right]_{L/8}^L$$

$$= \left(\frac{8M}{3L}\right)\left[\left(\frac{L}{8}\right)^3\right] + \left(\frac{8M}{21L}\right)\left[L^3 - \left(\frac{L}{8}\right)^3\right] = \frac{ML^2}{192} + \left(\frac{8M}{21L}\right)\left[\frac{511L^3}{512}\right]$$

$$= \frac{ML^2}{192} + \left(\frac{M}{3}\right)\left[\frac{73L^2}{64}\right] = \frac{74ML^2}{192} = \boxed{0.385\ ML^2}$$

REFLECT

At first, it's difficult to see if this is a reasonable value or not. Let's consider a few similar cases for comparison. The moment of inertia for a uniform pencil of mass $2M$ and length L

rotating about an axis through one end would be $I = \dfrac{2ML^2}{3} = 0.667ML^2$. It makes sense that our answer should be *smaller* than this since the $2M$ is not distributed uniformly. The moment of inertia of an object made up of two rods—one of length $L/8$ and mass M, one of length $7L/8$ and mass M—about an axis through each of their ends would be $I = \dfrac{1}{3}M\left(\dfrac{L}{8}\right)^2 + \dfrac{1}{3}M\left(\dfrac{7L}{8}\right)^2 = \dfrac{ML^2}{192} + \dfrac{49ML^2}{192} = \dfrac{50ML^2}{192} = 0.26\ ML^2$. In this case, the mass is distributed closer to the rotation axis than in our pencil example. Therefore, we would expect our answer to be *larger* than this moment of inertia. Comparing similar cases is a quick and useful way of determining whether an answer, especially an algebraic one, is reasonable or not.

8.77

SET UP

A bowling ball of known mass $m = 5$ kg and radius $R = 11$ cm is rolling without slipping down a lane at an angular speed of $\omega = 2\dfrac{\text{rad}}{\text{s}}$. We need to determine the ratio of the ball's translational kinetic energy $\left(K_{\text{tr}} = \dfrac{1}{2}mv_{\text{CM}}^2\right)$ to its rotational kinetic energy $\left(K_{\text{rot}} = \dfrac{1}{2}I\omega^2\right)$. The translational speed of the ball is related to the angular speed by $v_{\text{CM}} = R\omega$. The ball is a sphere, so its moment of inertia is $I = \dfrac{2}{5}mR^2$.

SOLVE

$$\frac{K_{\text{tr}}}{K_{\text{rot}}} = \frac{\left(\dfrac{1}{2}mv_{\text{CM}}^2\right)}{\left(\dfrac{1}{2}I\omega^2\right)} = \frac{m(R\omega)^2}{\left(\dfrac{2}{5}mR^2\right)\omega^2} = \boxed{\dfrac{5}{2}}$$

REFLECT

The solution is *independent* of the mass of the ball, the size of the ball, or the speed the ball is rolling!

8.81

SET UP

A Frisbee of mass $m = 0.160$ kg and diameter $d = 0.25$ m is spinning at an angular speed of 300 rpm about an axis through its center. First we are asked to find the rotational kinetic energy of the Frisbee. We are given the angular speed of the Frisbee but not the moment of inertia. The mass is *not* uniformly distributed in this case—70% of the mass can be modeled as a thin ring of diameter d, while the remaining 30% of the mass can be modeled as a flat disk of diameter d. The moment of inertia of the Frisbee is the sum of the moments of inertia for the ring and the disk. Knowing that only 60% of the

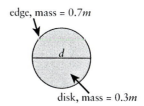

Figure 8-7 Problem 81

rotational kinetic energy is converted into gravitational potential energy, we can solve for the maximum height the Frisbee attains above the release point.

SOLVE

Kinetic energy of the Frisbee:

$$I_{Frisbee} = I_{ring} + I_{disk} = m_{ring}r^2_{ring} + \frac{1}{2}m_{disk}r^2_{disk} = (0.7m)\left(\frac{d}{2}\right)^2 + \frac{1}{2}(0.3m)\left(\frac{d}{2}\right)^2$$

$$= (0.7)(0.160 \text{ kg})\left(\frac{0.25 \text{ m}}{2}\right)^2 + \frac{1}{2}(0.3)(0.160 \text{ kg})\left(\frac{0.25 \text{ m}}{2}\right)^2 = 0.002125 \text{ kg} \cdot \text{m}^2$$

$$K_{rot} = \frac{1}{2}I\omega^2 = \frac{1}{2}(0.002125 \text{ kg} \cdot \text{m}^2)\left(300\frac{\text{rev}}{\text{min}} \times \frac{1 \text{ min}}{60 \text{ s}} \times \frac{2\pi \text{ rad}}{1 \text{ rev}}\right)^2 = 1.05 \text{ J}$$

Maximum height:

$$0.6K_{rot} = mgh$$

$$h = \frac{0.6K_{rot}}{mg} = \frac{0.6(1.05 \text{ J})}{(0.16 \text{ kg})\left(9.81\frac{\text{m}}{\text{s}^2}\right)} = \boxed{0.401 \text{ m}}$$

REFLECT

This works out to be about 1.3 feet, which is reasonable. If the energy conversion were 100% efficient, the maximum height would have been 0.669 m or 2.2 ft.

8.87

SET UP

A CD player initially rotates at 300 rpm. It speeds up to 450 rpm within 0.75 s. We will assume that the CD player undergoes constant acceleration from 300 rpm to 450 rpm and we will use rotational kinematics. We will need to convert from rpm to rad/s.

SOLVE

$$450\frac{\text{rev}}{\text{min}} \times \frac{1 \text{ min}}{60 \text{ s}} \times \frac{2\pi \text{ rad}}{1 \text{ rev}} = 15\pi\frac{\text{rad}}{\text{s}}$$

$$300\frac{\text{rev}}{\text{min}} \times \frac{1 \text{ min}}{60 \text{ s}} \times \frac{2\pi \text{ rad}}{1 \text{ rev}} = 10\pi\frac{\text{rad}}{\text{s}}$$

$$\omega_f = \omega_i + \alpha t$$

$$\alpha = \frac{\omega_f - \omega_i}{t} = \frac{\left(15\pi\frac{\text{rad}}{\text{s}}\right) - \left(10\pi\frac{\text{rad}}{\text{s}}\right)}{0.75 \text{ s}} = \frac{20\pi}{3}\frac{\text{rad}}{\text{s}^2} = \boxed{21\frac{\text{rad}}{\text{s}^2}}$$

REFLECT

Although this seems large, it is reasonable given that the increase of 150 rpm happens in under a second. Remember to convert rpm into rad/s using 1 minute = 60 seconds and 1 revolution = 2π radians.

Get Help: P'Cast 8.3 – A Stopping Top

8.93

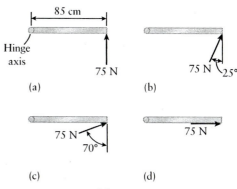

(a)   (b)

(c)   (d)

Figure 8-8 Problem 93

SET UP

A 75-N force acts on an 85-cm-wide door at various angles in parts (a)–(d). We are asked to calculate the torque that this force exerts about an axis through the hinges in each case. Because the force is exerted in the same spot on the door, the magnitude of $\vec{r}$ is the same in each case, namely, 85 cm. The angle θ between $\vec{r}$ and $\vec{F}$ changes, though.

SOLVE

Part a) $\theta = 90°$, so $\tau = RF \sin(\theta) = (0.85 \text{ m})(75 \text{ N}) \sin(90°) = \boxed{63.8 \text{ N} \cdot \text{m}}$.

Part b) $\theta = 115°$, so $\tau = RF \sin(\theta) = (0.85 \text{ m})(75 \text{ N}) \sin(115°) = \boxed{57.8 \text{ N} \cdot \text{m}}$.

Part c) $\theta = 160°$, so $\tau = RF \sin(\theta) = (0.85 \text{ m})(75 \text{ N}) \sin(160°) = \boxed{21.8 \text{ N} \cdot \text{m}}$.

Part d) $\theta = 0°$, so $\tau = RF \sin(\theta) = (0.85 \text{ m})(75 \text{ N}) \sin(0°) = \boxed{0}$.

REFLECT

The maximum torque should occur when the force is applied perpendicularly to the door (part a), and the minimum torque should occur when the force is applied parallel to the door (part d). For angles in between, we expect the torque to be an intermediate value.

Get Help: Interactive Example – Pivot Rod with Weight

8.97

SET UP

A person is holding a barbell of mass $m = 10$ kg in a completely outstretched hand of length $L = 0.75$ m. We're asked to calculate the magnitude of the torque due to the barbell about the shoulder, $\tau = RF\sin(\theta)$. The force of interest is the force of the barbell on the hand, which acts downward. We need to draw a free-body diagram of the barbell at rest and use Newton's second law in order to determine the magnitude of this force. The force of the arm on the barbell is related to the force of the barbell on the arm through Newton's third law. The distance from the rotation axis to the point where the force is applied is the length of the arm, L. Since the force of the barbell on the arm acts straight down and the arm is completely outstretched, the angle between the force vector and the $\vec{r}$ vector is 90°.

SOLVE

Free-body diagram of the barbell and Newton's second law:

$$\sum F_y = F_{\text{arm} \rightarrow \text{barbell}} - F_g = ma_y = 0$$

$$F_{\text{arm} \rightarrow \text{barbell}} = F_g = mg$$

Newton's third law:

$$F_{\text{arm} \rightarrow \text{barbell}} = F_{\text{barbell} \rightarrow \text{arm}} = mg$$

Figure 8-9 Problem 97

Solving for the magnitude of the torque,

$$\tau = RF\sin(\theta) = Lmg\sin(90°) = (0.75 \text{ m})(10 \text{ kg})\left(9.8\frac{\text{m}}{\text{s}^2}\right) = \boxed{73.5 \text{ N} \cdot \text{m}}$$

REFLECT

Since we're interested in the forces acting on the arm, we use the contact force of the barbell on the arm, not the *weight* of the barbell, in calculating the torque. The weight of the barbell is the force on the barbell due to the Earth, which would not appear on a free-body diagram of the arm and is *not* the third law partner of the contact force. The two forces are related, though; we just need to draw a free-body diagram for the barbell at rest and solve Newton's second law.

8.103

SET UP

We are asked to find the magnitude of the angular momentum ($L = I\omega$) of a 70-kg person riding in a Ferris wheel of diameter $d = 35$ m that makes a complete revolution in 25 s. The person can be treated as a point mass rotating at a distance $d/2$ (that is, the radius) about the center of the Ferris wheel. We can calculate the angular speed of the person by knowing that 1 revolution is 2π radians and assuming the angular speed is constant.

SOLVE

$$I_{\text{point mass}} = mr^2 = m\left(\frac{d}{2}\right)^2 = (70 \text{ kg})\left(\frac{35 \text{ m}}{2}\right)^2 = 21{,}438 \text{ kg} \cdot \text{m}^2$$

$$\omega = \frac{\Delta\theta}{\Delta t} = \frac{2\pi \text{ rad}}{25 \text{ s}} = 0.251\frac{\text{rad}}{\text{s}}$$

$$L = I\omega = (21{,}438 \text{ kg} \cdot \text{m}^2)\left(0.251\frac{\text{rad}}{\text{s}}\right) = \boxed{5388\frac{\text{kg} \cdot \text{m}^2}{\text{s}}}$$

REFLECT

Recall that r for the moment of inertia of a point mass is the distance the mass is located from the axis of rotation. In this case it is equal to the radius of the Ferris wheel since the passenger cars are located along the rim.

Get Help: P'Cast 8.5 – Figure Skater

8.109

SET UP

The first derivative of the angular momentum vector with respect to time is the torque vector. Since we are given the torque vector as a function of time, we need to integrate it with respect to time to get the angular momentum vector. Assuming the magnitude of the angular momentum vector is zero at $t = 0$ allows us to evaluate a definite integral from $t = 0$ up to some arbitrary t. This will give us the angular momentum vector as a function of time. Once we solve for the angular momentum as a function of time, we can plug in $t = 0.5$ s into the expression to get the angular momentum vector at that time and determine its magnitude. For ease of calculation, we'll assume that the torque vector is given in terms of SI units.

SOLVE

$$\vec{\tau} = \frac{d\vec{L}}{dt}$$

$$\vec{L}(t) = \int_{L(0)}^{L(t)} d\vec{L} = \int_0^t \vec{\tau} dt = \int_0^t (3\cos(\pi t)\hat{x} + 4\cos(\pi t)\hat{y}) dt$$

$$= \left[\frac{3}{\pi}\sin(\pi t)\hat{x} + \frac{4}{\pi}\sin(\pi t)\hat{y}\right]_0^t = \frac{3}{\pi}\sin(\pi t)\hat{x} + \frac{4}{\pi}\sin(\pi t)\hat{y}$$

$$\vec{L}(t = 0.5 \text{ s}) = \frac{3}{\pi}\sin\left(\frac{\pi}{2}\right)\hat{x} + \frac{4}{\pi}\sin\left(\frac{\pi}{2}\right)\hat{y} = \frac{3}{\pi}\hat{x} + \frac{4}{\pi}\hat{y}$$

$$L = \sqrt{L_x^2 + L_y^2} = \sqrt{\left(\frac{3}{\pi}\right)^2 + \left(\frac{4}{\pi}\right)^2} = \frac{1}{\pi}\sqrt{(3)^2 + (4)^2} = \boxed{\frac{5}{\pi}\frac{\text{kg} \cdot \text{m}^2}{\text{s}} = 1.59\frac{\text{kg} \cdot \text{m}^2}{\text{s}}}$$

REFLECT

The torque vector is only a function of one variable (t) even though it has two components (x and y); $\hat{x}$ and $\hat{y}$ are unit vectors, not variables. This allows for a straightforward integration of each component.

8.117

SET UP

The bike rim is to be made up of a hoop of mass $m_{\text{hoop}} = 1$ kg and radius $R = 0.5$ m and an unknown number of spokes, each of mass $m_{\text{spoke}} = 0.01$ kg. Since the spokes connect the center of a typical bicycle wheel to the hoop, the length of each one is equal to the radius of the hoop. The overall moment of inertia of the bike rim rotating about an axis through its center is equal to the moment of inertia of the hoop plus the moment of inertia for *each* spoke. We can model each spoke as a thin rod of length R rotating about an axis through one end. From this we can calculate the number of spokes necessary to give the stated overall moment of inertia. Once we know the number of spokes, we can determine the mass of all of the spokes and add that to the mass of the hoop to get the overall mass of the rim.

SOLVE

Part a)

$$I_{\text{rim}} = I_{\text{hoop}} + NI_{\text{spoke}} = m_{\text{hoop}}R^2 + N\left(\frac{1}{3}m_{\text{spoke}}L^2\right) = m_{\text{hoop}}R^2 + N\left(\frac{1}{3}m_{\text{spoke}}R^2\right)$$

$$N = (I_{\text{rim}} - m_{\text{hoop}}R^2)\left(\frac{3}{m_{\text{spoke}}R^2}\right)$$

$$= ((0.280 \text{ kg} \cdot \text{m}^2) - (1 \text{ kg})(0.5 \text{ m})^2)\left(\frac{3}{(0.01 \text{ kg})(0.5 \text{ m})^2}\right) = \boxed{36}$$

Part b)

$$m_{\text{rim}} = m_{\text{hoop}} + Nm_{\text{spoke}} = (1 \text{ kg}) + 36(0.01 \text{ kg}) = \boxed{1.36 \text{ kg}}$$

REFLECT

This is a reasonable mass for a bike rim. We aren't explicitly given the length of the spokes, but from our everyday knowledge about bicycle wheels we are able to get started on the problem. Picturing how a typical bike wheel is constructed and how it rotates helps determine which moment of inertia formulas are appropriate to use. Using everyday knowledge to make reasonable assumptions is an invaluable skill in solving physics problems.

8.125

SET UP

When the space debris accumulates on the surface of the Earth, it changes both the mass and moment of inertia of the Earth, which is why we expect the Earth's angular speed to decrease. We can use conservation of angular momentum if we treat the space debris and the Earth as an isolated system. Our initial state will be the Earth spinning at its "normal" rate (1 revolution in 24 hr) as the space debris falls radially inward from space. We assume the debris falls radially inward until it hits the Earth's surface because this simplifies the calculation since the initial angular momentum of the space debris will be zero, $\vec{L} = \vec{r} \times \vec{p} = \vec{r} \times (-\hat{r}) = 0$. The final state will be the Earth and space debris spinning together. We will assume the debris is uniformly distributed on the surface of the Earth in a thin layer. This allows us to model the final moment of inertia as a thin, spherical shell with the radius of the Earth plus the moment of inertia of the Earth itself, $I_f = I_{Earth} + I_{debris} = I_{solid\ sphere} + I_{spherical\ shell}$. Part (a) of the question asks for the *change* in the angular speed per year, so we'll be solving for $\Delta\omega = \omega_f - \omega_i$. Once we know $\Delta\omega$ for one year we can use the definition of the period to determine the total change in angular speed if the Earth's rotation period changed by 1 s. Comparing these two values of $\Delta\omega$ will tell us how many years of accumulating debris it would take for that change to occur.

SOLVE

Part a)

$$L_i = L_f$$

$$I_i\omega_i = I_f\omega_f$$

$$\omega_f = \frac{I_i\omega_i}{I_f}$$

$$\Delta\omega = \omega_f - \omega_i = \left(\frac{I_i\omega_i}{I_f}\right) - \omega_i = \omega_i\left(\frac{I_i}{I_f} - 1\right) = \omega_i\left(\frac{\left(\frac{2}{5}m_E R_E^2\right)}{\left(\frac{2}{5}m_E R_E^2\right) + \left(\frac{2}{3}m_{debris} R_E^2\right)} - 1\right)$$

$$= \omega_i\left(\frac{m_E}{m_E + \left(\frac{5}{3}m_{debris}\right)} - 1\right) = \omega_i\left(\frac{m_E}{m_E + \left(\frac{5}{3}m_{debris}\right)} - \frac{m_E + \left(\frac{5}{3}m_{debris}\right)}{m_E + \left(\frac{5}{3}m_{debris}\right)}\right)$$

$$= -\omega_i\left(\frac{\left(\frac{5}{3}m_{debris}\right)}{m_E + \left(\frac{5}{3}m_{debris}\right)}\right)$$

Converting the angular speed and mass of the debris:

$$\omega_i = \frac{2\pi \text{ rad}}{24 \text{ hr}} \times \frac{1 \text{ hr}}{3600 \text{ s}} = 7.3 \times 10^{-5}\frac{\text{rad}}{\text{s}}$$

$$m_{\text{debris}} = 60{,}000 \text{ tons} \times \frac{2000 \text{ lb}}{1 \text{ ton}} \times \frac{1 \text{ kg}}{2.2 \text{ lb}} = 5.5 \times 10^7 \text{ kg}$$

Plugging in:

$$\Delta\omega = -\left(7.3 \times 10^{-5}\frac{\text{rad}}{\text{s}}\right)\left(\frac{\frac{5}{3}(5.5 \times 10^7 \text{ kg})}{(5.97 \times 10^{24} \text{ kg}) + \frac{5}{3}(5.5 \times 10^7 \text{ kg})}\right) = \boxed{-1.1 \times 10^{-21}\frac{\text{rad}}{\text{s}}}$$

This is the change in angular speed in one year.

Part b)

$$T = \frac{2\pi}{\omega}$$

$$\Delta T = \frac{2\pi}{\omega_f} - \frac{2\pi}{\omega_i} = \frac{2\pi}{\omega_i + \Delta\omega} - \frac{2\pi}{\omega_i} = 2\pi\left(\frac{1}{\omega_i + \Delta\omega} - \frac{1}{\omega_i}\right)$$

$$\frac{\Delta T}{2\pi} + \frac{1}{\omega_i} = \frac{1}{\omega_i + \Delta\omega}$$

$$\omega_i + \Delta\omega = \frac{1}{\left(\frac{\Delta T}{2\pi} + \frac{1}{\omega_i}\right)} = \frac{1}{\left(\frac{\omega_i(\Delta T) + 2\pi}{2\pi\omega_i}\right)} = \frac{2\pi\omega_i}{\omega_i(\Delta T) + 2\pi}$$

$$\Delta\omega = \frac{2\pi\omega_i}{\omega_i(\Delta T) + 2\pi} - \omega_i = \omega_i\left(\frac{2\pi}{\omega_i(\Delta T) + 2\pi} - 1\right)$$

$$= \left(7.3 \times 10^{-5}\frac{\text{rad}}{\text{s}}\right)\left(\frac{2\pi}{\left(7.3 \times 10^{-5}\frac{\text{rad}}{\text{s}}\right)(1 \text{ s}) + 2\pi} - 1\right)$$

$$= -8.5 \times 10^{-10}\frac{\text{rad}}{\text{s}}$$

This is the overall change in angular speed. To find how long it takes to reach this overall change in angular speed, we need to divide it by the change in angular speed *per year*:

$$\text{time} = \frac{\left(-8.5 \times 10^{-10}\frac{\text{rad}}{\text{s}}\right)}{\left(-1 \times 10^{-21}\frac{\text{rad}}{\text{s} \cdot \text{yr}}\right)} = \boxed{7.7 \times 10^{11} \text{ yr}}$$

REFLECT

Although 60,000 tons of debris sounds like a lot, it is negligible compared to the mass of the Earth. The amount of time we calculated for the Earth's rotation to change by a full second is longer than the age of the Earth (approximately 4×10^9 years), so we don't have to worry about space debris changing the Earth's rotation significantly...which is good.

Get Help: P'Cast 8.5 – Figure Skater

8.131

SET UP

A marble of mass M, radius R, and moment of inertia

$I = \dfrac{2}{5}MR^2$ is placed in front of a spring with spring constant

k that has been compressed an amount x_c. The spring is released and, when the spring reaches its equilibrium length, the marble

Figure 8-10 Problem 131

comes off the spring and begins to roll without slipping. The statement that static friction does no work suggests that we use conservation of mechanical energy to determine the time t it takes for the marble to travel a distance D. The initial elastic potential energy stored in the spring is converted into both translational and rotational kinetic energy of the marble. By solving for the linear speed of the marble as it leaves the spring, we can then use the definition of speed to determine the time it takes to travel a distance D. In part (b) we can use dimensional analysis to ensure our algebraic answer from part (a) has the correct dimension of time.

SOLVE

Part a)

Conservation of mechanical energy:

$$U_{\text{elastic}} = K_{\text{rot}} + K_{\text{tr}}$$

$$\frac{1}{2}kx_c^2 = \frac{1}{2}I\omega^2 + \frac{1}{2}Mv^2$$

$$\frac{1}{2}kx_c^2 = \frac{1}{2}\left(\frac{2}{5}MR^2\right)\left(\frac{v}{R}\right)^2 + \frac{1}{2}Mv^2$$

$$\frac{1}{2}kx_c^2 = \frac{1}{5}Mv^2 + \frac{1}{2}Mv^2 = \frac{7}{10}Mv^2$$

$$v = \sqrt{\frac{5kx_c^2}{7M}} = \frac{D}{t}$$

$$t = \frac{D}{v} = \frac{D}{\sqrt{\dfrac{5kx_c^2}{7M}}} = D\sqrt{\frac{7M}{5kx_c^2}} = \boxed{\frac{D}{x_c}\sqrt{\frac{7M}{5k}}}$$

Part b)

$$[t] = [\mathrm{T}]$$

$$[M] = [\mathrm{M}]$$

$$[D] = [\mathrm{L}]$$

$$[x_c] = [\mathrm{L}]$$

$$[k] = [\mathrm{M}][\mathrm{T}]^{-2}$$

$$[t] \stackrel{?}{=} \frac{[D]}{[x_c]}\sqrt{\frac{[M]}{[k]}}$$

$$[\mathrm{T}] = \frac{[\mathrm{L}]}{[\mathrm{L}]}\sqrt{\frac{[\mathrm{M}]}{[\mathrm{M}][\mathrm{T}]^{-2}}} = \sqrt{[\mathrm{T}]^2} = [\mathrm{T}]$$

REFLECT

Because the work done by all the nonconservative forces on the marble is equal to zero, we could invoke the conservation of mechanical energy because $W_{\text{nonconservative}} = \Delta E_{\text{mechanical}}$. We also explicitly use dimensional analysis in this problem. It's always a good idea to check the dimensions of your final answer whether or not it is required to do so.

Chapter 9
Elasticity and Fracture

Conceptual Questions

9.3 Yes. Real cables have weight, and that weight can be sufficient to break the cable.

9.7 The strain is the more fundamental physical property. The strain determines the force the material is applying or its stress. There is no guarantee that each stress occurs at exactly one strain. In particular, very stretched metal begins to lose strength, and so the stress decreases.

9.11 $\kappa = -\dfrac{1}{V}\dfrac{dV}{dP}; \kappa = -\dfrac{1}{V}\dfrac{\Delta V}{\Delta P}.$

9.15 The shear modulus is a measure of how much an object will deform under a given stress. The larger the shear modulus, the less it will deform; it is more rigid. So, the term *rigidity* applies because a material will deform less if it has a larger shear modulus.

Multiple-Choice Questions

9.23 B ($\sqrt{2}$). The applied force, Young's modulus, and original length of the cable remain constant:

$$A_1(\Delta L)_1 = A_2(\Delta L)_2$$

$$\pi\left(\frac{D_1}{2}\right)^2(\Delta L) = \pi\left(\frac{D_2}{2}\right)^2\left(\frac{\Delta L}{2}\right)$$

$$D_2 = D_1\sqrt{2}$$

Get Help: P'Cast 9.1 – ACLs and Steel

9.27 A (materials with a relatively large bulk modulus). Usually, you would not want the volume of the building to change much when a force acts on it.

Estimation Questions

9.31 10×10^9 N/m^2 (approximately the same as for oak wood).

Get Help: P'Cast 9.1 – ACLs and Steel

9.37

Strain (%)	Stress (10^9 N/m^2)
0.0	0
0.1	125
0.2	250
0.3	230
0.4	230
0.5	235
0.6	240
0.7	250
0.8	260
0.9	270
1.0	300
1.5	325
2.0	350
2.5	375
3.0	400
3.5	375
4.0	350
4.5	325
5.0	300

Figure 9-2 Problem 57

Part a) The yield strength is the tensile stress at which the material becomes permanently deformed: 250×10^9 N/m^2.

Part b) The ultimate strength is the maximum tensile stress the material can withstand: 400×10^9 N/m^2.

Part c) The Young's modulus is the slope of the initial linear region: $125{,}000 \times 10^9$ N/m^2.

Part d) Point of rupture occurs when the stress of an object is so large that the material starts to lose its structural integrity: 300×10^9 N/m^2. (This is where the graph stops.)

Problems

9.41

SET UP

An aluminum bar has a cross section of 1 cm^2 and an initial length of 88 cm. An applied force stretches the bar to a new length of 100 cm. We can calculate the magnitude of the force from the Young's modulus of aluminum and the tensile strain in the bar from the definition. The Young's modulus of aluminum from Table 9-1 is 70×10^9 N/m^2.

SOLVE

$$\frac{F}{A} = Y\frac{\Delta L}{L_0}$$

$$F = YA\frac{\Delta L}{L_0} = \left(70 \times 10^9 \frac{\text{N}}{\text{m}^2}\right)(10^{-4}\ \text{m}^2)\left(\frac{(100\ \text{cm}) - (88\ \text{cm})}{88\ \text{cm}}\right) = \boxed{9.55 \times 10^5\ \text{N} = 955\ \text{kN}}$$

$$\frac{\Delta L}{L_0} = \frac{(100\ \text{cm}) - (88\ \text{cm})}{88\ \text{cm}} = \boxed{0.136 = 13.6\%}$$

REFLECT
We would expect the force required to stretch an aluminum bar to be large.

Get Help: P'Cast 9.1 – ACLs and Steel

9.45

SET UP
We can calculate the ratio of the tensile strains of an aluminum bar and a steel bar from their respective Young's moduli because the force and cross-sectional area are constant in both cases. The Young's modulus of steel is 200×10^9 N/m^2; the Young's modulus of aluminum is 70×10^9 N/m^2. The original length of each bar does not affect the strain since the strain is essentially the percent change in the object's length.

SOLVE

Part a)

$$\frac{\left(\dfrac{\Delta L}{L_0}\right)_{Al}}{\left(\dfrac{\Delta L}{L_0}\right)_{Steel}} = \frac{\left(\dfrac{F}{AY}\right)_{Al}}{\left(\dfrac{F}{AY}\right)_{Steel}} = \frac{Y_{Steel}}{Y_{Al}} = \frac{\left(200\dfrac{N}{m^2}\right)}{\left(70\dfrac{N}{m^2}\right)} = \frac{20}{7} = \boxed{2.86}$$

Part b) The strain does not depend on the relative lengths of the bars (but the relative amount of compression does).

REFLECT
It makes sense that aluminum, with a smaller Young's modulus, should experience a larger strain than steel for a given force.

Get Help: P'Cast 9.1 – ACLs and Steel

9.47

SET UP
A steel piano wire has a diameter of 0.2 cm. The tensile strength, which is the maximum tensile stress the material can withstand, of steel is 5.0×10^8 N/m^2. We can calculate the magnitude of the tension required to break the cylindrical wire directly from the strength of steel.

SOLVE

$$5.0 \times 10^8 \frac{N}{m^2} = \left(\frac{F}{A}\right)_{max} = \frac{F_{max}}{\pi R^2}$$

$$F_{max} = \left(5.0 \times 10^8 \frac{N}{m^2}\right)(\pi R^2) = \left(5.0 \times 10^8 \frac{N}{m^2}\right)(\pi)\left(\frac{0.20 \times 10^{-2}\ m}{2}\right)^2 = \boxed{1600\ N = 1.6\ kN}$$

REFLECT
The original length of the wire is irrelevant in this problem.

Get Help: P'Cast 9.1 – ACLs and Steel

9.53

SET UP
A rigid cube of volume V_0 is filled with water and then frozen solid. The water expands by 9%, which means the volume changes by $+0.09V_0$. We can calculate the pressure exerted on the sides of the cube from the bulk modulus. The bulk modulus of water is 2.0×10^9 N/m^2.

SOLVE

$$\frac{F}{A} = B\frac{\Delta V}{V_0} = B\frac{0.09 V_0}{V_0} = \left(2.0 \times 10^9 \frac{\text{N}}{\text{m}^2}\right)(0.09) = \boxed{1.8 \times 10^8 \frac{\text{N}}{\text{m}^2}}$$

REFLECT

We ignored the negative sign in the bulk modulus equation because we were just interested in the magnitude of the pressure. If we included the negative sign, we would need to determine the pressure required to squeeze the ice back into the original cube.

9.57

SET UP

A force of 200,000 N is acting tangentially to the top face of a brass plate. The top face is 2 cm by 20 cm. We can calculate the shear strain from the shear stress and the shear modulus. The shear modulus of brass from Table 9-1 is 40×10^9 N/m^2. The shear strain is equal to the tangent of φ.

Figure 9-2 Problem 57

SOLVE

$$\frac{F}{A} = S\frac{x}{h}$$

$$\frac{x}{h} = \frac{F}{AS} = \frac{200,000 \text{ N}}{(0.02 \text{ m})(0.20 \text{ m})\left(40 \times 10^9 \frac{\text{N}}{\text{m}^2}\right)} = \boxed{0.00125}$$

$$\tan(\varphi) = \frac{x}{h}$$

$$\varphi = \arctan\left(\frac{x}{h}\right) = \arctan(0.0025) = \boxed{0.072°}$$

REFLECT

The top face will hang over the bottom face by 125 microns.

Get Help: P'Cast 9.4 – Earthquake Damage

9.61

SET UP

In parts of the cardiovascular system, the cells lining the walls of the blood vessels experience a shear stress of 20 dyne/cm^2. We are told the shear strain is 0.008. We can find the shear modulus directly from these data. We will need the conversion factor from dynes to newtons: $1 \text{ N} = 10^5$ dyne.

SOLVE

$$S = \frac{\left(\frac{F}{A}\right)}{\left(\frac{x}{h}\right)} = \frac{\left(20\frac{\text{dyne}}{\text{cm}^2}\right)}{0.008} \times \frac{1 \text{ N}}{10^5 \text{ dyne}} \times \frac{10^4 \text{ cm}^2}{1 \text{ m}^2} = \boxed{250\frac{\text{N}}{\text{m}^2}}$$

REFLECT

We expect cells to have a much smaller shear modulus than, say, metal.

Get Help: P'Cast 9.4 – Earthquake Damage

9.65

SET UP

A given rope can support 10,000 N/m² before breaking. A knot is tied in it, which decreases the ultimate breaking strength by a factor of 2. A worker wants to use this knotted rope to hang a load of 1000 N/m² but needs to make sure it meets the safety standards set by his company. The company uses a safety factor of 10 for all of its ropes, which means the ultimate breaking strength of the rope must be at least 10 times larger than the load it is supporting. To determine the safety factor for using this rope to hold this load, we need to compare the maximum force the rope can support to the force of the load. If this is more than 10, then the worker can safely use the rope; if not, he'll need to find a new, stronger rope.

SOLVE

Part a)

The overestimated ultimate breaking strength of the knotted rope is

$$\frac{\left(10{,}000\dfrac{\text{N}}{\text{m}^2}\right)}{2} = 5000\dfrac{\text{N}}{\text{m}^2}$$

The safety factor in this case is

$$\frac{\left(5000\dfrac{\text{N}}{\text{m}^2}\right)}{\left(1000\dfrac{\text{N}}{\text{m}^2}\right)} = \boxed{5}$$

Part b) $\boxed{\text{No}}$, the rigging company cannot use this rope because the safety factor is less than 10.

REFLECT

Since the cross section of the rope remains constant, the ratio between the ultimate breaking strength and the applied stress is equal to the ratio between the maximum force and the applied force.

Get Help: P'Cast 9.6 – ACL Breaking Strain

9.69

SET UP

Strands of caterpillar silk are typically 2.0 μm in diameter and have a Young's modulus of 4.0×10^9 N/m². We are asked to consider a climbing rope made of silk strands that is initially 9.0 m long and only stretches a distance of 1.00 cm when supporting the weight of two 85-kg climbers. We can calculate the cross sectional area of this rope from the weight, the Young's modulus, and the strain. The ratio of this cross-sectional area to the cross-sectional area of one strand will give us the number of strands in the rope. Assuming the strands are distributed in a cylinder, we can use the area of a circle to calculate the diameter of the rope.

SOLVE

Part a)

$$\frac{F}{A} = Y\frac{\Delta L}{L_0}$$

$$A = \frac{FL_0}{Y\Delta L} = \frac{2(85 \text{ kg})\left(9.8\frac{\text{m}}{\text{s}^2}\right)(9.0 \text{ m})}{\left(4 \times 10^9 \frac{\text{N}}{\text{m}^2}\right)(0.0100 \text{ m})} = 0.000375 \text{ m}^2$$

$$N_{\text{strands}} = \frac{A_{\text{total}}}{A_{1 \text{ strand}}} = \frac{0.000375 \text{ m}^2}{\pi\left(\frac{2.0 \times 10^{-6} \text{ m}}{2}\right)^2} = \boxed{1.2 \times 10^8 \text{ strands}}$$

Part b)

$$A_{\text{total}} = \pi\left(\frac{D}{2}\right)^2$$

$$D = \sqrt{\frac{4A_{\text{total}}}{\pi}} = \sqrt{\frac{4(0.000375 \text{ m}^2)}{\pi}} = \boxed{0.022 \text{ m} = 2.2 \text{ cm}}$$

Yes, this a reasonably-sized rope for mountain climbers to carry.

REFLECT

Climbing ropes made out of nylon have a comparable Young's modulus, 3.0×10^9 N/m².

Get Help: P'Cast 9.1 – ACLs and Steel

9.73

SET UP

A weightlifter lifts a mass of 263.5 kg above his head. All of this weight is supported by the lifter's two tibias. The average length of a tibia is 385 mm and its diameter is 3.0 cm. The Young's modulus for bone is 2.0×10^{10} N/m². We can use these data to calculate the distance his tibia is compressed upon lifting the weight. In order to determine if that distance is significant, we need to compare it to the initial length of the tibia. Throughout the calculation, we will assume that the given length of the tibia (385 mm) includes compression due to his weight.

SOLVE

Part a)

$$\frac{F}{A} = Y\frac{\Delta L}{L_0}$$

$$\Delta L = \frac{FL_0}{AY} = \frac{(263.5 \text{ kg})\left(9.8\frac{\text{m}}{\text{s}^2}\right)(385 \text{ mm})}{2(\pi)\left(\frac{0.03 \text{ m}}{2}\right)^2\left(2.0 \times 10^{10}\frac{\text{N}}{\text{m}^2}\right)} = \boxed{0.035 \text{ mm}}$$

Part b) No, a change of 0.035 mm is not a significant compression.

Part c) No, we do not need to include his weight because we are interested in the compression due to extra weight. His tibia is initially compressed slightly due to his weight; we've assumed the value of 385 mm includes that initial compression.

REFLECT

If the value of 385 mm did *not* include the initial compression of his weight, his weight would compress the tibia by 0.02 mm, which is also negligible.

9.79

SET UP

On Earth, the spherical bubbles near the surface of a glass of water have a diameter $D_0 = 2.5$ mm. This glass of water is taken to Mars. The atmospheric pressure on Earth is 1.0×10^5 N/m^2, and the atmospheric pressure on Mars is 650 N/m^2. We can relate the change in the diameter and, thus, the volume to the change in atmospheric pressure through the bulk modulus.

SOLVE

$$\Delta P = -B \frac{\Delta V}{V_0}$$

$$\Delta P = -B \frac{\left(\frac{4}{3}\pi R_f^3\right) - \left(\frac{4}{3}\pi R_0^3\right)}{\left(\frac{4}{3}\pi R_0^3\right)} = -B\left(\frac{R_f^3}{R_0^3} - 1\right)$$

$$R_f = R_0 \sqrt[3]{1 - \frac{\Delta P}{B}}$$

$$D_f = D_0 \sqrt[3]{1 - \frac{\Delta P}{B}} = (2.5 \text{ mm})\sqrt[3]{1 - \frac{\left(\left(650 \frac{\text{N}}{\text{m}^2}\right) - \left(1.0 \times 10^5 \frac{\text{N}}{\text{m}^2}\right)\right)}{\left(1.01 \times 10^5 \frac{\text{N}}{\text{m}^2}\right)}} = \boxed{3.14 \text{ mm}}$$

REFLECT

The change in pressure is approximately 10^5 as is the bulk modulus of air. Therefore, the percent change in volume is about 100%, which means the volume on Mars is about $2^{1/3}$ that on Earth, or $(2^{1/3})(2.5 \text{ mm}) = 3.15$ mm.

9.85

SET UP

A rectangular object made from a material that has a Young's modulus of Y and a bulk modulus of B has dimensions of ℓ, w, and t. A stress acts along ℓ, which changes the dimensions of the object to $(\ell + \Delta \ell)$, $(w + \Delta w)$, and $(t + \Delta t)$, where $\Delta \ell < 0$, $\Delta w > 0$, and $\Delta t > 0$. We can find an expression for the fractional change in volume $\frac{\Delta V}{V_0}$ in terms of these dimensions and relate it to the bulk modulus through $\Delta P = -B\frac{\Delta V}{V_0}$. We will need to invoke the definition of Poisson's ratio ν, which is the induced strain divided by the primary strain: $\nu = -\frac{\Delta w/w}{\Delta \ell/\ell} = -\frac{\Delta t/t}{\Delta \ell/\ell}$. The fact that the fluid exerts a pressure, which is the primary applied stress, allows us to connect the bulk modulus to the Young's modulus.

Figure 9-3 Problem 85

SOLVE

$$V_0 = \ell w t$$

$$V_f = (\ell + \Delta\ell)(w + \Delta w)(t + \Delta t)$$

$$\Delta V = V_f - V_0 = (\ell + \Delta\ell)(w + \Delta w)(t + \Delta t) - \ell w t$$

$$\frac{\Delta V}{V_0} = \frac{(\ell + \Delta\ell)(w + \Delta w)(t + \Delta t) - \ell w t}{\ell w t} = \left(\frac{\ell + \Delta\ell}{\ell}\right)\left(\frac{w + \Delta w}{w}\right)\left(\frac{t + \Delta t}{t}\right) - 1$$

$$= \left(1 + \frac{\Delta\ell}{\ell}\right)\left(1 + \frac{\Delta w}{w}\right)\left(1 + \frac{\Delta t}{t}\right) - 1$$

But $\nu = -\dfrac{\Delta w/w}{\Delta\ell/\ell} = -\dfrac{\Delta t/t}{\Delta\ell/\ell}$, which means

$$\frac{\Delta w}{w} = -\nu\left(\frac{\Delta\ell}{\ell}\right) = -\nu\varepsilon \quad \text{and} \quad \frac{\Delta t}{t} = -\nu\left(\frac{\Delta\ell}{\ell}\right) = -\nu\varepsilon$$

Plugging this into our expression for $\dfrac{\Delta V}{V_0}$:

$$\frac{\Delta V}{V_0} = \left(1 + \frac{\Delta\ell}{\ell}\right)\left(1 + \frac{\Delta w}{w}\right)\left(1 + \frac{\Delta t}{t}\right) - 1 = (1 + \varepsilon)(1 - \nu\varepsilon)(1 - \nu\varepsilon) - 1$$

$$= (1 + \varepsilon)(1 - 2\nu\varepsilon + \nu^2\varepsilon^2) - 1 = 1 + \varepsilon - 2\nu\varepsilon - 2\nu\varepsilon^2 + \nu^2\varepsilon^2 + \nu^2\varepsilon^3 - 1$$

We can neglect all terms with ε^2 or higher since $\varepsilon \ll 1$:

$$\frac{\Delta V}{V_0} \approx \varepsilon - 2\nu\varepsilon = \varepsilon(1 - 2\nu)$$

When this object is immersed in a fluid and symmetrically stressed, this strain will need to be multiplied by 3 for the three dimensions:

$$\frac{\Delta V}{V_0} \approx 3\varepsilon(1 - 2\nu)$$

Setting the pressure change equal to the stress applied to the object:

$$\Delta P = \left(0 - \frac{F}{A}\right) = -\frac{F}{A}$$

$$-B\frac{\Delta V}{V_0} = -Y\frac{\Delta\ell}{\ell}$$

$$B(3\varepsilon(1 - 2\nu)) = Y\varepsilon$$

$$\boxed{B = \frac{Y}{3(1 - 2\nu)}}$$

REFLECT

For an object under compression, the strain will always be less than 1 since the final length is smaller than the initial length. We are assuming the fluid does not change the overall length by very much.

Chapter 10
Gravitation

Conceptual Questions

10.1 Newton reasoned that the Moon, as it orbits, falls back toward the center of Earth due to the pull of gravity. Incrementally, it will move tangent to the orbital path, and then it will fall back to Earth. The combined motion leads to the familiar circular path. An apple falls straight toward Earth without the tangential motion. The vertical motion is the same for both the apple and the Moon.

10.3 When the mass of one object is doubled, the force between two objects doubles. If both masses are halved, the force is one-fourth the original value.

Get Help: P'Cast 10.1 – The Force between Earth and You

10.13 By Kepler's law, the faster an object is moving as it orbits the Sun, the closer it is to the Sun. Thus, Earth is closer to the Sun when the Northern Hemisphere is in winter.

Get Help: P'Cast 10.6 – How Long is a Year

10.17 Cells are probably the least affected by gravity (owing to strong electromagnetic forces at that level), but tissues and organs certainly depend on the constant supply of blood. This supply is intrinsically tied to the gravimetric forces on fluids in our body. Our vertebrae, for example, are continually compressed when we live on Earth. Our feet become flatter due to our continuous weight pushing down on them; our muscles sag over time, and so on.

Multiple-Choice Questions

10.21 A (she weighs more at the bottom). The value of g decreases with increasing altitude because the distance between the Earth and the mountain climber is increasing.

10.23 B ($\sqrt{2}v$). We can calculate the speed of the satellite in the initial orbit using Newton's second law: $v = \sqrt{\dfrac{GM}{R}}$. Comparing this to the escape speed formula, we see that the escape speed must be $\sqrt{2}v$.

10.25 A (Satellite A has more kinetic energy, less potential energy, and less mechanical energy (potential energy plus kinetic energy) than satellite B). The kinetic energy of an orbiting satellite is inversely proportional to the distance. Less potential energy in this case means "more negative."

Estimation/Numerical Analysis

10.29 Assuming it is possible to measure a force of about 10^{-7} N, the weights should have about 1 mm center-to-center separation.

     Get Help: P'Cast 10.1 – The Force between Earth and You

10.33 The force between Earth and Jupiter is 1/18,000 of the force between Earth and the Sun. The impact would be small.

     Get Help: P'Cast 10.1 – The Force between Earth and You

Problems

10.37

SET UP

A baseball ($m_{\text{ball}} = 0.150$ kg) is 100 m from a bat ($m_{\text{bat}} = 0.935$ kg). We can calculate the magnitude of the gravitational force between the two from Newton's universal law of gravitation.

SOLVE

$$F_{\text{g}} = \frac{Gm_{\text{ball}}m_{\text{bat}}}{r^2} = \frac{\left(6.67 \times 10^{-11}\dfrac{\text{N} \cdot \text{m}^2}{\text{kg}^2}\right)(0.150 \text{ kg})(0.935 \text{ kg})}{(1000 \text{ m})^2} = \boxed{9.35 \times 10^{-16} \text{ N}}$$

REFLECT

As expected, this force is negligible.

     Get Help: P'Cast 10.1 – The Force between Earth and You

10.41

SET UP

We can use Newton's universal law of gravitation to calculate the net force acting on the Moon during an eclipse. The net force is equal to the vector sum of the gravitational force due to the Sun and the gravitational force due to the Earth. In our coordinate system, we'll assume that all three satellites lie along the x-axis and that the Sun is to the left of the Moon (toward $-x$).

SOLVE

$$\sum F_x = F_{\text{S}\rightarrow\text{M},\, x} + F_{\text{E}\rightarrow\text{M},\, x} = -\frac{Gm_{\text{S}}m_{\text{M}}}{r_{\text{SM}}^2} + \frac{Gm_{\text{E}}m_{\text{M}}}{r_{\text{EM}}^2} = GM_{\text{M}}\left(\frac{m_{\text{E}}}{r_{\text{EM}}^2} - \frac{m_{\text{S}}}{(r_{\text{ES}} - r_{\text{EM}})^2}\right)$$

$$= \left(6.67 \times 10^{-11}\frac{\text{N} \cdot \text{m}^2}{\text{kg}^2}\right)(7.35 \times 10^{22} \text{ kg})$$

$$\left(\frac{5.98 \times 10^{24} \text{ kg}}{3.84 \times 10^8 \text{ m}^2} - \frac{1.99 \times 10^{30} \text{ kg}}{((1.50 \times 10^{11} \text{ m}) - (3.84 \times 10^8 \text{ m}))^2}\right)$$

$$= -2.37 \times 10^{20} \text{ N}$$

The net force acting on the Moon has a magnitude of $\boxed{2.37 \times 10^{20} \text{ N and points toward the Sun}}$.

Get Help: P'Cast 10.1 – The Force between Earth and You

REFLECT
The gravitational force on the Moon due to the Sun is still larger than the force due to the Earth even though the Earth is much closer.

10.43

SET UP
We are interested in the point where the net gravitational force due to the Earth ($m_{\text{E}} = 5.98 \times 10^{24}$ kg) and the Moon ($m_{\text{M}} = 7.35 \times 10^{22}$ kg) on a space probe (mass m_{p}) is equal to zero; the point where the net force is equal to zero must be between the Earth and the Moon. The Moon is 3.84×10^8 m from the Earth. We will let x be the distance from the Earth to the probe and y be the distance from the Moon to the probe. We'll assume that the Earth is to the left of the probe and the Moon is to the right.

SOLVE

$$\frac{GM_{\text{E}}M_{\text{p}}}{x^2} = \frac{GM_{\text{M}}M_{\text{p}}}{y^2}$$

$$\frac{x^2}{y^2} = \frac{M_{\text{E}}}{M_{\text{M}}}$$

$$x = y\sqrt{\frac{M_{\text{E}}}{M_{\text{M}}}}$$

$$x + y = 3.84 \times 10^8 \text{ m}$$

$$y\sqrt{\frac{M_{\text{E}}}{M_{\text{M}}}} + y = 3.84 \times 10^8 \text{ m}$$

$$y = \frac{3.84 \times 10^8 \text{ m}}{1 + \sqrt{\dfrac{M_{\text{E}}}{M_{\text{M}}}}} = \frac{3.84 \times 10^8 \text{ m}}{1 + \sqrt{\dfrac{5.98 \times 10^{24} \text{ kg}}{7.35 \times 10^{22} \text{ kg}}}} = 3.83 \times 10^7 \text{ m}$$

$$x = (3.84 \times 10^8 \text{ m}) - y = 3.46 \times 10^8 \text{ m}$$

The point where the net gravitational force due to the Earth and Moon is equal to zero is $\boxed{3.46 \times 10^8 \text{ m to the right of the Earth}}$ or $\boxed{3.83 \times 10^7 \text{ m to the left of the Moon}}$.

REFLECT
The exact mass of the space probe is not necessary to solve this problem. It makes sense that the point where the net force is zero should be closer to the Moon than the Earth since it is less massive.

Get Help: P'Cast 10.1 – The Force between Earth and You

10.47

SET UP

A hollow sphere has a mass m_1 and radius r_1. A second mass m_2 is located a distance r_2 from the surface of the sphere. We can use Newton's universal law of gravitation to calculate the magnitude of the gravitational force between the two objects. Mass m_2 is located outside of the hollow sphere, which means the sphere acts as a point mass located a distance $(r_1 + r_2)$ from m_2 due to the shell theorem.

Figure 10-1 Problem 47

SOLVE

$$F_g = \frac{Gm_1m_2}{(r_1 + r_2)^2}$$

REFLECT

The gravitational force between the two objects is attractive and points along the line connecting the two.

10.53

SET UP

Five spherical storage tanks are separated by 75 m and filled with natural gas ($\rho_{gas} = 0.8$ kg/m³). Each storage tank stores a volume of 5000 m³; we can use the inner volume of the tank to find the radius of the inside of each tank. The tanks are constructed from a 10-cm-thick wall of steel, which has a density of $\rho_{steel} = 8000$ kg/m³. The distance d between the centers of neighboring tanks is equal to twice the radius of the tank plus twice the thickness of the steel wall plus the 75 m separating the tanks. The total mass of each tank is the mass of the steel plus the mass of the gas. We can use Newton's universal law of gravitation to calculate the gravitational force on the end tank, which we'll call tank 1, by the other tanks. The net gravitational force will point toward the remaining tanks.

SOLVE

Figure 10-2 Problem 53

Radius of each tank:

$$\frac{4}{3}\pi R^3 = 5000 \text{ m}^3$$

$$R = \sqrt[3]{\frac{3(5000 \text{ m}^3)}{4\pi}} = \sqrt[3]{\frac{3(5000 \text{ m}^3)}{4\pi}} = 10.61 \text{ m}$$

Distance between the center of mass of the tanks:

$$d = 2(10.61 \text{ m}) + 2(0.1 \text{ m}) + (75 \text{ m}) = 96.42 \text{ m}$$

Mass of each tank with gas:

$$M = M_{\text{steel}} + M_{\text{gas}} = \left[\frac{4}{3}\pi(R + (0.1 \text{ m}))^3 - \frac{4}{3}\pi R^3\right]\rho_{\text{steel}} + (5000 \text{ m}^3)\rho_{\text{gas}}$$

$$= \frac{4}{3}\pi[((10.61 \text{ m}) + (0.1 \text{ m}))^3 - (10.61 \text{ m})^3]\left(8000\frac{\text{kg}}{\text{m}^3}\right) + (5000 \text{ m}^3)\left(0.8\frac{\text{kg}}{\text{m}^3}\right)$$

$$= (1.142 \times 10^6 \text{ kg}) + (4000 \text{ kg}) = 1.146 \times 10^6 \text{ kg}$$

Force of tank 2 on tank 1:

$$F_{2\to1} = \frac{Gm_2m_1}{d_{21}^2} = \frac{\left(6.67 \times 10^{-11}\frac{\text{N} \cdot \text{m}^2}{\text{kg}^2}\right)(1.146 \times 10^6 \text{ kg})^2}{(96.42 \text{ m})^2} = 0.00942 \text{ N}$$

Force of tank 3 on tank 1:

$$F_{3\to1} = \frac{Gm_3m_1}{d_{31}^2} = \frac{\left(6.67 \times 10^{-11}\frac{\text{N} \cdot \text{m}^2}{\text{kg}^2}\right)(1.146 \times 10^6 \text{ kg})^2}{[2(96.42 \text{ m})]^2} = 0.00236 \text{ N}$$

Force of tank 4 on tank 1:

$$F_{4\to1} = \frac{Gm_4m_1}{d_{41}^2} = \frac{\left(6.67 \times 10^{-11}\frac{\text{N} \cdot \text{m}^2}{\text{kg}^2}\right)(1.146 \times 10^6 \text{ kg})^2}{[3(96.42 \text{ m})]^2} = 0.00105 \text{ N}$$

Force of tank 5 on tank 1:

$$F_{5\to1} = \frac{Gm_5m_1}{d_{51}^2} = \frac{\left(6.67 \times 10^{-11}\frac{\text{N} \cdot \text{m}^2}{\text{kg}^2}\right)(1.146 \times 10^6 \text{ kg})^2}{[4(96.42 \text{ m})]^2} = 0.000589 \text{ N}$$

Net force on tank 1:

$$\sum F = (0.00942 \text{ N}) + (0.00236 \text{ N}) + (0.00105 \text{ N}) + (0.000589 \text{ N}) = 0.0134 \text{ N}$$

The net gravitational force acting on tank 1 has a magnitude of

$\boxed{0.0134 \text{ N and points toward the other tanks}}$.

REFLECT

The weight of tank 1 is approximately 1×10^7 N, which is nine orders of magnitude larger than the net gravitational force due to the other four tanks.

10.55

SET UP

A 10-kg object moves from a point 6000 m above sea level to a point 1000 m above sea level. The work done by gravity is equal to the negative of the change in the gravitational potential energy of the object. We need to include the radius of the Earth when determining the initial and final positions due to the shell theorem.

SOLVE

$$W = -\Delta U = -\left(\frac{-Gm_E m}{R_f} - \left(\frac{-Gm_E m}{R_i}\right)\right) = Gm_E m\left(\frac{1}{R_E + (1000 \text{ m})} - \frac{1}{R_E + (6000 \text{ m})}\right)$$

$$= \left(6.67 \times 10^{-11}\frac{\text{N} \cdot \text{m}^2}{\text{kg}^2}\right)(5.98 \times 10^{24} \text{ kg})(10 \text{ kg})$$

$$\left(\frac{1}{(6.38 \times 10^6 \text{ m}) + (1000 \text{ m})} - \frac{1}{(6.38 \times 10^6 \text{ m}) + (6000 \text{ m})}\right)$$

$$= \boxed{489,400 \text{ J} = 489.4 \text{ kJ}}$$

REFLECT

The force of gravity and the displacement are in the same direction, so we expect the work done by gravity to be positive. If we used $U = mgy$ instead, the work done by gravity would be 490 kJ.

10.59

SET UP

The volume of water in the Pacific Ocean is 7×10^8 km^3. The density of seawater is 1030 kg/m^3. The mass of the Moon is 7.35×10^{22} kg, and the distance from the center of the Moon to the center of the Earth is 3.84×10^8 m. We can use these data to calculate the gravitational potential energy between the Pacific Ocean and the Moon when the Pacific is facing away and facing toward the Moon. Although it is a very rough approximation, we will treat the ocean as a point mass. Therefore, we will need to add and subtract, respectively, the radius of the Earth in each calculation. The maximum speed of the water due to the tidal influence of the Moon can be calculated through conservation of energy—the change in the potential energies calculated in parts (a) and (b) is equal to the change in kinetic energy.

SOLVE

$$m_w = \rho_w V = \left(1030\frac{\text{kg}}{\text{m}^3}\right)\left(7 \times 10^8 \text{ km}^3 \times \left(\frac{10^3 \text{ m}}{1 \text{ km}}\right)^3\right) = 7.21 \times 10^{20} \text{ kg}$$

Part a)

$$U = -\frac{Gm_w m_M}{r_{away}} = -\frac{\left(6.67 \times 10^{-11}\frac{\text{N} \cdot \text{m}^2}{\text{kg}^2}\right)(7.21 \times 10^{20} \text{ kg})(7.35 \times 10^{22} \text{ kg})}{(3.84 \times 10^8 \text{ m}) + (6.38 \times 10^6 \text{ m})}$$

$$= \boxed{-9.054 \times 10^{24} \text{ J}}$$

Part b)

$$U = -\frac{Gm_w m_M}{r_{toward}} = -\frac{\left(6.67 \times 10^{-11}\frac{\text{N} \cdot \text{m}^2}{\text{kg}^2}\right)(7.21 \times 10^{20} \text{ kg})(7.35 \times 10^{22} \text{ kg})}{(3.84 \times 10^8 \text{ m}) - (6.38 \times 10^6 \text{ m})} = \boxed{-9.360 \times 10^{24} \text{ J}}$$

Part c)

$$\Delta K = -\Delta U$$

$$\frac{1}{2}m_\text{w}v^2 = -((-9.360 \times 10^{24}\text{ J}) - (-9.054 \times 10^{24}\text{ J})) = 3.06 \times 10^{23}\text{ J}$$

$$v = \sqrt{\frac{2(3.06 \times 10^{23}\text{ J})}{m_\text{w}}} = \sqrt{\frac{2(3.06 \times 10^{23}\text{ J})}{7.21 \times 10^{20}\text{ kg}}} = \boxed{29\frac{\text{m}}{\text{s}}}$$

REFLECT

This is a very rough approximation and overestimation, as it treats the Pacific Ocean as a point mass and doesn't take into account drag.

10.61

SET UP

We can use conservation of energy to calculate the impact speed of a 100-kg asteroid that is initially moving at a speed of 200 m/s at a distance of 1000 km from the Moon's surface. When calculating the initial and final gravitational potential energies of the asteroid, we need to include the radius of the Moon ($r_\text{M} = 1.737 \times 10^6$ m). The force of the Moon on the asteroid is nonconservative and the only force doing work to stop the asteroid. Therefore, the work done by the Moon on the asteroid is equal to the change in the asteroid's mechanical energy. We will assume that the asteroid does not move very deeply into the surface during the impact, so $\Delta U \approx 0$.

SOLVE

Part a)

$$U_\text{i} + K_\text{i} = U_\text{f} + K_\text{f}$$

$$-\frac{Gm_\text{M}m}{r_\text{i}} + \frac{1}{2}mv_\text{i}^2 = -\frac{Gm_\text{M}m}{r_\text{f}} + \frac{1}{2}mv_\text{f}^2$$

$$v_\text{f} = \sqrt{-\frac{2Gm_\text{M}}{r_\text{i}} + \frac{2Gm_\text{M}}{r_\text{f}} + v_\text{i}^2} = \sqrt{-2Gm_\text{M}\left(\frac{1}{(r_\text{M} + (10^6\text{ m}))} - \frac{1}{r_\text{M}}\right) + v_\text{i}^2}$$

$$= \sqrt{-2\left(6.67 \times 10^{-11}\frac{\text{N} \cdot \text{m}^2}{\text{kg}^2}\right)(7.35 \times 10^{22}\text{ kg})\left(\frac{1}{((1.737 \times 10^6\text{ m}) + (10^6\text{ m}))} - \frac{1}{(1.737 \times 10^6\text{ m})}\right) + \left(200\frac{\text{m}}{\text{s}}\right)^2}$$

$$= \boxed{1450\frac{\text{m}}{\text{s}}}$$

Part b)

$$W_\text{nc} = \Delta K + \Delta U = \frac{1}{2}m(v_\text{f}^2 - v_\text{i}^2) + 0 = \frac{1}{2}m(0 - v_\text{i}^2) = \frac{1}{2}(100\text{ kg})\left(1450\frac{\text{m}}{\text{s}}\right)^2 = -1.05 \times 10^8\text{ J}$$

The Moon does $\boxed{1.05 \times 10^8\text{ J}}$ of work.

REFLECT

We would expect the asteroid to heat up and/or break apart upon impact as well as embed itself into the Moon's surface, so our calculation gives an upper bound to the work done by the Moon.

10.65

SET UP

A satellite orbits the Earth once every 86.5 min. Assuming its orbit is circular, we can use Kepler's law of periods to calculate the radius of the orbit, and then the circumference of the satellite's orbit, from the period.

SOLVE

$$T^2 = \frac{4\pi^2 R^3}{GM}$$

$$R = \sqrt[3]{\frac{GMT^2}{4\pi^2}} = \sqrt[3]{\frac{\left(6.67 \times 10^{-11}\frac{N \cdot m^2}{kg^2}\right)(5.98 \times 10^{24} \text{ kg})\left(86.5 \text{ min} \times \frac{60 \text{ s}}{1 \text{ min}}\right)^2}{4\pi^2}} = 6.48 \times 10^6 \text{ m}$$

$$C = 2\pi R = 2\pi(6.48 \times 10^6 \text{ m}) = \boxed{4.07 \times 10^7 \text{ m}}$$

REFLECT

Be sure to use a consistent set of units when performing your calculations.

Get Help: P'Cast 10.7 – Satellites

10.71

SET UP

The Moon orbits the Earth in a nearly circular orbit once every 27.32 days. We can use Kepler's law of periods to help calculate the distance d from the surface of the Moon to the surface of Earth. The radius in Kepler's law is the center-to-center distance between the Moon and the Earth, so we will need to subtract out the radius of the Earth and the radius of the Moon in order to find d.

SOLVE

Converting the period into seconds:

$$27.32 \text{ days} \times \frac{24 \text{ hr}}{1 \text{ day}} \times \frac{3600 \text{ s}}{1 \text{ hr}} = 2.360 \times 10^6 \text{ s}$$

Calculating the distance:

$$T^2 = \frac{4\pi^2 R^3}{GM} = \frac{4\pi^2(R_{\text{Earth}} + R_{\text{Moon}} + d)^3}{GM}$$

$$d = \sqrt[3]{\frac{GMT^2}{4\pi^2}} - R_{\text{Earth}} - R_{\text{Moon}} = \sqrt[3]{\frac{\left(6.67 \times 10^{-11}\frac{N \cdot m^2}{kg^2}\right)(5.98 \times 10^{24} \text{ kg})(2.360 \times 10^6 \text{ s})^2}{4\pi^2}}$$

$$- (6.38 \times 10^6 \text{ m}) - (1.73 \times 10^6 \text{ m})$$

$$= \boxed{3.751 \times 10^8 \text{ m}}$$

REFLECT

The mean distance from the center of the Earth to the center of the Moon is 3.84×10^8 m. This is larger than the distance between their surfaces, which makes sense.

Get Help: P'Cast 10.7 – Satellites

10.73

SET UP

A thin, uniform rod with a mass M and a length L is located on the x-axis. A small object of mass m is located a distance D away along the x-axis at point P. We can calculate the gravitational force of the rod on the small object by integrating the infinitesimal gravitational field due to each portion of the rod and then multiplying by the mass m. Through symmetry, we see that the y component of the field (and force) is equal to zero, so we only need to calculate the x component. The x component of the infinitesimal field dg_x is equal to the value of the field $dg = \dfrac{Gdm}{r^2}$, where $r = (D + x)$ is the distance from the point P to the location of the mass dm. We are told that the rod is uniform, which means we can use the linear mass density to convert from an integral with respect to mass to an integral with respect to x.

Figure 10-3 Problem 73

SOLVE

$$g_x = \int -dg_x = -\int \frac{Gdm}{r^2}$$

But $dm = \dfrac{M}{L}dx$:

$$g_x = -\int \frac{Gdm}{r^2} = -\int_0^L \frac{G}{(D+x)^2}\frac{M}{L}dx = \frac{GM}{L}\int_L^0 \frac{1}{(D+x)^2}dx = \frac{GM}{L}\left[-\frac{1}{D+x}\right]_L^0$$

$$= -\frac{GM}{L}\left[\frac{1}{D} - \frac{1}{D+L}\right]$$

Therefore, the force exerted on a small object at point P is:

$$\vec{F} = -\frac{GMm}{L}\left[\frac{1}{D} - \frac{1}{D+L}\right]\hat{x} = -\frac{GMm}{L}\left[\frac{(D+L) - D}{D(D+L)}\right]\hat{x} = \boxed{-\frac{GMm}{D(D+L)}\hat{x}}$$

REFLECT

Thinking about the situation beforehand and invoking symmetry will usually save you some time and effort when solving a problem. The gravitational force is always attractive, so the force on the small object must point toward $-x$. If $D \gg L$, the expression for the force reduces to $\vec{F} = -\dfrac{GMm}{D^2}\hat{x}$; in other words, from far away, the rod looks like a point mass.

10.77

SET UP

A shuttle is launched into a low-Earth orbit $(R \approx R_E)$ near the equator. We can use conservation of energy to calculate the necessary launch speed of the shuttle v_L. A shuttle launched due east is launched with the rotation of the Earth, while a shuttle launched due west is launched against the rotation of the Earth, so the rotation speed of the Earth will either be added or subtracted to the launch speed v_L of the rocket. U.S. space shuttles were launched from the east coast of Florida because it is close to the equator and an eastward launch would fly directly over the ocean.

SOLVE

Tangential speed of the Earth's surface:

$$v_E = \frac{2\pi R_E}{T} = \frac{2\pi(6.38 \times 10^6 \text{ m})}{\left(24 \text{ hr} \times \dfrac{3600 \text{ s}}{1 \text{ hr}}\right)} = 464\frac{\text{m}}{\text{s}}$$

Part a)

$$K_i + U_i = K_f + U_f$$

$$\frac{1}{2}m\left(v_L + \left(464\frac{\text{m}}{\text{s}}\right)\right)^2 - \frac{GM_E m}{R_E} = 0$$

$$v_L = \sqrt{\frac{2GM_E}{R_E}} - \left(464\frac{\text{m}}{\text{s}}\right) = \sqrt{\frac{2\left(6.67 \times 10^{-11}\dfrac{\text{N} \cdot \text{m}^2}{\text{kg}^2}\right)(5.98 \times 10^{24} \text{ kg})}{(6.38 \times 10^6 \text{ m})}} - \left(464\frac{\text{m}}{\text{s}}\right)$$

$$= \boxed{10{,}720\frac{\text{m}}{\text{s}} = 10.72\frac{\text{km}}{\text{s}}}$$

Part b)

$$K_i + U_i = K_f + U_f$$

$$\frac{1}{2}m\left(v_L - \left(464\frac{\text{m}}{\text{s}}\right)\right)^2 - \frac{GM_E m}{R_E} = 0$$

$$v_L = \sqrt{\frac{2GM_E}{R_E}} + \left(464\frac{\text{m}}{\text{s}}\right) = \sqrt{\frac{2\left(6.67 \times 10^{-11}\dfrac{\text{N} \cdot \text{m}^2}{\text{kg}^2}\right)(5.98 \times 10^{24} \text{ kg})}{(6.38 \times 10^6 \text{ m})}} + \left(464\frac{\text{m}}{\text{s}}\right)$$

$$= \boxed{11{,}650\frac{\text{m}}{\text{s}} = 11.65\frac{\text{km}}{\text{s}}}$$

Part c) U.S. space missions are launched from the east coast of Florida so that the launch site will be as close to the equator as possible while still being on the U.S. mainland. Also, an eastward launch from its location won't pass over inhabited areas.

REFLECT

It makes sense that the space shuttle needs to launch with a smaller speed if it launches with the rotation of the Earth.

10.81

SET UP

The semimajor axis of an elliptical orbit of a planet around a star is $a = 2.25$ AU and the semiminor axis is $b = 1.75$ AU. The eccentricity e of an ellipse is defined as $e = \sqrt{1 - \dfrac{b^2}{a^2}}$. The perihelion r_p and aphelion r_a are the closest and farthest distances from the star to the planet; we can also represent the eccentricity, as well as the semimajor axis, in terms of these distances. Finally, the area of an ellipse is $A = \pi ab$.

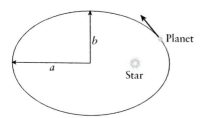

Figure 10-4 Problem 81

SOLVE

Eccentricity:

$$e = \sqrt{1 - \frac{b^2}{a^2}} = \sqrt{1 - \frac{(1.75 \text{ AU})^2}{(2.25 \text{ AU})^2}} = \boxed{0.63}$$

Perihelion and aphelion:

$$e = \frac{r_a - r_p}{r_a + r_p}$$

$$er_a + er_p = r_a - r_p$$

$$r_a = \left(\frac{1 + e}{1 - e}\right)r_p$$

$$2a = r_a + r_p = \left(\frac{1 + e}{1 - e}\right)r_p + r_p = \left(\frac{1 + e + 1 - e}{1 - e}\right)r_p = \left(\frac{2}{1 - e}\right)r_p$$

$$r_p = (1 - e)a = (1 - (0.63))(2.25 \text{ AU}) = \boxed{0.833 \text{ AU}}$$

$$r_a = 2a - r_p = 2(2.25 \text{ AU}) - (0.833 \text{ AU}) = \boxed{3.67 \text{ AU}}$$

Area:

$$A = \pi ab = \pi(2.25 \text{ AU})(1.75 \text{ AU}) = \boxed{12.4 \text{ AU}^2} \times \frac{(1.496 \times 10^{11} \text{ m})^2}{1 \text{ AU}^2} = \boxed{2.78 \times 10^{23} \text{ m}^2}$$

REFLECT

The planet moves the fastest at the perihelion and the slowest at the aphelion.

10.89

SET UP

Two objects ($m_{10} = 10$ kg and $m_3 = 3$ kg) are separated by 40 cm. A third object ($m_1 = 1$ kg) is placed at a location along the line connecting them such that the net force acting on m_1 is zero. By considering the force vectors, this location must be between the two original objects. We will define x as the distance between m_{10} and m_1 and y as the distance between m_3 and m_1. Setting the magnitudes of the gravitational forces equal and using the fact that $x + y = 40$ cm, we can solve for x.

Figure 10-5 Problem 89

SOLVE

$$\frac{Gm_{10}m_1}{x^2} = \frac{Gm_3m_1}{y^2}$$

$$x^2 = \frac{m_{10}y^2}{m_3}$$

$$x = y\sqrt{\frac{m_{10}}{m_3}}$$

But $x + y = 40$ cm:

$$x + y = y\sqrt{\frac{m_{10}}{m_3}} + y = y\sqrt{\frac{10 \text{ kg}}{3 \text{ kg}}} + y = 2.826y = 40 \text{ cm}$$

$$y = \frac{40 \text{ cm}}{2.826} = 14.2 \text{ cm}$$

$$x = (40 \text{ cm}) - y = (40 \text{ cm}) - (14.2 \text{ cm}) = \boxed{25.8 \text{ cm}}$$

REFLECT

It makes sense that the 1-kg object needs to be closer to the less massive object if the net force acting on it is zero.

Get Help: P'Cast 10.1 – The Force between Earth and You

Chapter 11
Fluids

Conceptual Questions

11.3 No. The pressure on the water is determined by its depth, gravity, and the pressure pushing down on the surface (air pressure). The ability of the dam to withstand that pressure does, of course, depend on its shape.

11.9 It is easier to float in the Great Salt Lake because salt water is denser than freshwater.

11.15 The wind flow over the roof is very fast, while the airflow inside the house is not; this lowers the exterior pressure. If the roof is not made to hold down against a large part of an atmosphere of pressure pushing it up, it will fly off.

Multiple-Choice Questions

11.19 D (be increased, but not necessarily doubled). Because the gauge pressure is the pressure above 1 atm, doubling the gauge pressure will not necessarily double the absolute pressure.

11.23 A (one-quarter the speed in the 0.5-cm pipe).

$$v_1 A_1 = v_2 A_2$$

$$v_2 = v_1 \frac{A_1}{A_2} = v_1 \left(\frac{\pi (0.5 \text{ cm})^2}{\pi (1 \text{ cm})^2} \right) = \boxed{\frac{1}{4} v_1}$$

Get Help: P'Cast 11.8 – Blood Flow
P'Cast 11.9 – From Capillaries to the Vena Cavae

11.27 A (Box A).

$$\sum F_y = F_b - F_g = \rho_{\text{water}} V_{\text{object}} g - \rho_{\text{object}} V_{\text{object}} g = m a_y = \rho_{\text{object}} V_{\text{object}} a_y$$

$$a_y = \frac{\rho_{\text{water}} V_{\text{object}} g - \rho_{\text{object}} V_{\text{object}} g}{\rho_{\text{object}} V_{\text{object}}} = g \left(\frac{\rho_{\text{water}}}{\rho_{\text{object}}} - 1 \right)$$

The object with the smaller density will experience a larger acceleration upward.

Get Help: Picture It – Buoyancy Force
Interactive Exercise – Bobber

Estimation Questions

11.31 Mercury is, of course, 760 mm; water is 13.6 times greater, or 10.4 m; and seawater is 10.1 m.

11.35 A wind that is 33 m/s (barely hurricane level) passing over the top of a 1-m² prone body resting on top of still air results in a lift of approximately 0.67 kN.

Problems

11.39

SET UP

A cylinder made of an unknown material is 20 cm long, 1 cm in radius, and has a mass of 490 g. We can calculate the average density of the material by dividing the cylinder's mass by its volume. The volume of a cylinder is the cross-sectional area multiplied by the length. We can compare our answer to part (a) to the densities in Table 11-1 in order to figure out the material from which the cylinder is made. The specific gravity of a material is equal to its density divided by the density of water, $\rho_{water} = 1000$ kg/m³.

SOLVE

$$\rho = \frac{M}{V} = \frac{M}{\pi R^2 L} = \frac{(0.490 \text{ kg})}{\pi (0.01 \text{ m})^2 (0.20 \text{ m})} = \boxed{7800 \frac{\text{kg}}{\text{m}^3}}$$

Part b)
The cylinder is most likely made of iron.

Part c)

$$SG = \frac{\rho}{\rho_{water}} = \frac{7800 \frac{\text{kg}}{\text{m}^3}}{1000 \frac{\text{kg}}{\text{m}^3}} = \boxed{7.80}$$

REFLECT

A "uniform" object has a constant density throughout it. If an object is not uniform, we can calculate the average density of the object as we did above.

Get Help: P'Cast 11.1 – Bony Chickens

11.45

SET UP

A neutron star has the same density as a neutron. We can calculate the radius of a neutron star that has the same mass as our Sun ($m_{star} = 1.99 \times 10^{30}$ kg) by setting the ratio of its mass to volume equal to the ratio of a neutron's mass to volume. We will assume both the star and the neutron are spheres. The mass of a neutron is $m_n = 1.7 \times 10^{-27}$ kg, and its approximate radius is 1.2×10^{-15} m.

SOLVE

$$\rho_n = \rho_{star}$$

$$\frac{m_n}{\left(\frac{4}{3}\pi R_n^3\right)} = \frac{m_{star}}{\left(\frac{4}{3}\pi R_{star}^3\right)}$$

$$R_{star} = R_n\sqrt[3]{\frac{m_{star}}{m_n}} = (1.2 \times 10^{-15}\text{ m})\sqrt[3]{\frac{1.99 \times 10^{30}\text{ kg}}{1.7 \times 10^{-27}\text{ kg}}} = \boxed{13\text{ km}}$$

REFLECT

A radius of 13 km seems reasonable for a "very small" star. Most astronomical distances are orders of magnitude larger. For instance, the radius of the Earth is approximately 6400 km and the distance between the Earth and the Moon is about 384,000 km.

Get Help: P'Cast 11.1 – Bony Chickens

11.49

SET UP

A force of 25 N is applied to the head of a nail that is 0.32 cm in diameter. The pointed end of the nail is 0.032 cm in diameter. The pressure on the head of the nail or the pointed end of the nail is equal to the applied force divided by the cross-sectional area at that location.

SOLVE

Part a)

$$P = \frac{F}{A} = \frac{F}{\pi\left(\frac{d}{2}\right)^2} = \frac{4F}{\pi d^2} = \frac{4(25\text{ N})}{\pi(0.32 \times 10^{-2}\text{ m})^2} = \boxed{3.11 \times 10^6\text{ Pa} = 3.11\text{ MPa}}$$

Part b)

$$P = \frac{F}{A} = \frac{F}{\pi\left(\frac{d}{2}\right)^2} = \frac{4F}{\pi d^2} = \frac{4(25\text{ N})}{\pi(0.032 \times 10^{-2}\text{ m})^2} = \boxed{3.11 \times 10^8\text{ Pa} = 311\text{ MPa}}$$

REFLECT

Since the pressure is inversely proportional to the area, a decrease of a factor of 10 in the radius corresponds to an increase by a factor of 100 in the pressure!

Get Help: P'Cast 11.3 – Submarine Hatch

11.53

SET UP

The difference in blood pressure between the top of the head and the bottom of the feet of a 1.75-m-tall person is equal to product of the density of blood ($\rho_{blood} = 1.06 \times 10^3\text{ kg/m}^3$), g, and the height. The conversion between pascals and mmHg is $1.01 \times 10^5\text{ Pa} = 760\text{ mmHg}$.

SOLVE

$$\Delta P = \rho_{\text{blood}} g h = \left(1.06 \times 10^3 \frac{\text{kg}}{\text{m}^3}\right)\left(9.8 \frac{\text{m}}{\text{s}^2}\right)(1.75 \text{ m}) = 18179 \text{ Pa} \times \frac{760 \text{ mmHg}}{1.01 \times 10^5 \text{ Pa}} = \boxed{137 \text{ mmHg}}.$$

REFLECT

Since we are just looking for the difference in blood pressure between the top of the head and the bottom of the feet, we do not need to explicitly include the pressure of the atmosphere in our calculation.

11.57

SET UP

Blood pressure is usually measured as a gauge pressure, which is the pressure above the atmospheric pressure. To convert a gauge pressure into an absolute pressure, we need to add the pressure of the atmosphere. In mmHg, the atmospheric pressure is 760 mmHg.

SOLVE

$$P = P_{\text{gauge}} + P_{\text{atm}} = (120 \text{ mmHg}) + (760 \text{ mmHg}) = \boxed{880 \text{ mmHg}}$$

REFLECT

Two common examples in which gauge pressures are used are for blood pressures and for car tires. Be sure to convert to absolute pressure by adding the atmospheric pressure before using the pressures in an equation.

11.61

SET UP

An airplane window has an area of 1000 cm². The pressure inside the airplane is 0.95 atm, while the pressure outside the airplane is 0.85 atm. The net force due to this pressure difference will point from the location of high pressure (inside the plane, in this case) toward the location of low pressure and have a magnitude equal to the pressure difference multiplied by the cross-sectional area of the window.

SOLVE

Pressure difference:

$$(0.95 \text{ atm}) - (0.85 \text{ atm}) = 0.10 \text{ atm} \times \frac{1.01 \times 10^5 \text{ Pa}}{1 \text{ atm}} = 1.01 \times 10^4 \text{ Pa}$$

Net force:

$$P = \frac{F}{A}$$

$$F = PA = (1.01 \times 10^4 \text{ Pa})\left(1000 \text{ cm}^2 \times \frac{1 \text{ m}^2}{10^4 \text{ cm}^2}\right) = 1.0 \times 10^3 \text{ N} = 1.0 \text{ kN}$$

The net force on the airplane window has a magnitude of $\boxed{1.0 \text{ kN and points outward}}$.

REFLECT

The net force due to a pressure difference always points from high pressure to low pressure.

11.65

SET UP

The hatch on the Mars lander is built and tested on Earth such that the internal pressure of the lander exactly balances the external pressure (that is, atmospheric pressure). The lander is then brought to Mars, where the external pressure is 650 Pa. The net force acting on the hatch is equal to the difference in the internal and external pressures multiplied by the cross-sectional area of the hatch. The hatch is round with a diameter of 0.500 m. Since the internal pressure of the lander is larger than the external pressure on Mars, the net force will point outward.

SOLVE

$$\Delta P = \frac{F}{A}$$

$$F = (\Delta P)A = ((1.01 \times 10^5 \text{ Pa}) - (650 \text{ Pa}))(\pi)\left(\frac{0.500 \text{ m}}{2}\right)^2$$

$$= \boxed{1.97 \times 10^4 \text{ N} \times \frac{0.22 \text{ lb}}{1 \text{ N}} = 4.33 \times 10^3 \text{ lb}}$$

This force points $\boxed{\text{outward}}$.

REFLECT

We can check our answer by looking at the orders of magnitude of each quantity:

$$F = (\Delta P)A \approx (10^5)\frac{(5 \times 10^{-1})^2}{4} = (10^5)\frac{(25)}{4}(10^{-2}) = (10^5)(10^1)(10^{-2}) = 10^4 \text{ N}$$

Get Help: P'Cast 11.3 – Submarine Hatch

11.69

SET UP

A hydraulic lift is made up of a small piston of area $A_1 = 0.033$ m^2 and a large piston of area $A_2 = 4$ m^2. An applied force of $F_1 = 16$ N causes the small piston to move downward a distance Δy_1. Pascal's principle lets us calculate the force that the large piston provides. The volume of the fluid moved on the left side must equal the volume of the fluid moved on the right side. We can find Δy_2 in terms of Δy_1 by setting these volumes equal to one another. The work done in moving each piston is equal to the magnitude of the force applied to the piston multiplied by the distance through which the piston travels.

Figure 11-1 Problem 69

SOLVE

Part a)

$$P_1 = P_2$$

$$\frac{F_1}{A_1} = \frac{F_2}{A_2}$$

$$F_2 = F_1 \frac{A_2}{A_1} = (16 \text{ N}) \frac{(4 \text{ m}^2)}{(0.033 \text{ m}^2)} = \boxed{1940 \text{ N}}$$

Part b)

$$A_1 \Delta y_1 = A_2 \Delta y_2$$

$$\Delta y_2 = \frac{A_1}{A_2} \Delta y_1 = \left(\frac{0.033 \text{ m}^2}{4 \text{ m}^2} \right) \Delta y_1 = \boxed{0.00825 \Delta y_1}$$

Part c)

$$W_1 = F_1 \Delta y_1 = (16 \text{ N})(0.20 \text{ m}) = \boxed{3.2 \text{ J}}$$

$$W_2 = F_2 \Delta y_2 = F_2(0.00825 \Delta y_1) = (1940 \text{ N})(0.00825)(0.20 \text{ m}) = \boxed{3.2 \text{ J}}$$

The work done in slowly pushing the small piston is $\boxed{\text{equal}}$ to the work done in raising the large piston.

REFLECT

Since the density and mass of the liquid inside the lift remain constant, the volume must also remain constant.

11.73

SET UP

A crown is weighed in air and in water. In air, the weight of the crown is measured as 5.15 N; we can calculate the mass of the crown from this measurement. In water, the apparent weight is 4.88 N. The apparent weight is equal to the weight of the crown in air minus the magnitude of the buoyant force, which is related to the volume of the crown. From both of these measurements, we can calculate the average density of the crown and compare it to the density of gold, which is 19.3 times the density of water. If the average density of the crown is close to the density of gold, then the crown is most likely made of gold.

SOLVE

$$F_g = m_{crown} g$$

$$m_{crown} = \frac{F_g}{g} = \frac{5.15 \text{ N}}{9.8 \frac{\text{m}}{\text{s}^2}} = 0.526 \text{ kg}$$

$$W_{app} = F_g - F_b = F_g - \rho_{water} g V_{crown}$$

$$V_{crown} = \frac{F_g - W_{app}}{\rho_{water}g} = \frac{(5.15 \text{ N}) - (4.88 \text{ N})}{\left(1000\frac{\text{kg}}{\text{m}^3}\right)\left(9.8\frac{\text{m}}{\text{s}^2}\right)} = 2.76 \times 10^{-5} \text{ m}^3$$

$$\rho_{crown} = \frac{m_{crown}}{V_{crown}} = \frac{0.526 \text{ kg}}{2.76 \times 10^{-5} \text{ m}^3} = 1.91 \times 10^4 \frac{\text{kg}}{\text{m}^3}$$

$$SG_{crown} = \frac{\rho_{crown}}{\rho_{water}} = \frac{1.91 \times 10^4 \frac{\text{kg}}{\text{m}^3}}{1000\frac{\text{kg}}{\text{m}^3}} = 19.1$$

This is close to the specific gravity of gold, so the $\boxed{\text{crown is most likely made of gold}}$.

REFLECT

The crown could still be fake, though. The density we calculated is the average density of the crown, which means it could be made up of a mixture of less dense and more dense materials.

Get Help: Picture It – Buoyancy Force
Interactive Exercise – Bobber

11.77

SET UP

A hose with a radius of 1 cm is connected to a faucet and used to fill a 5-L container in 45 s. The volume flow rate is equal to the total volume filled divided by the time it took to do so. The volume flow rate is also equal to the cross-sectional area of the hose multiplied by the speed of the water in the hose.

SOLVE

Part a)

$$Q = \frac{V}{t} = \frac{5 \text{ L}}{45 \text{ s}} \times \frac{1 \text{ m}^3}{1000 \text{ L}} = \boxed{1.11 \times 10^{-4}\frac{\text{m}^3}{\text{s}}}$$

Part b)

$$Q = Av = \pi R^2 v$$

$$v = \frac{Q}{\pi R^2} = \frac{1.111 \times 10^{-4}\frac{\text{m}^3}{\text{s}}}{\pi(1 \times 10^{-2} \text{ m}^2)^2} = \boxed{0.354\frac{\text{m}}{\text{s}}}$$

REFLECT

The volume flow rate will be constant since the hose is a closed system.

Get Help: P'Cast 11.8 – Blood Flow
P'Cast 11.9 – From Capillaries to the Vena Cavae

11.81

SET UP

The inner diameter of a needle used for an injection is 0.114×10^{-3} m. An injection of 2.5×10^{-3} L was given in 0.65 s. We can set the two expressions for the volume flow rate—the total volume divided by the time and the cross-sectional area of the needle multiplied by the speed of the fluid—equal and solve for the speed of the fluid as it leaves the needle.

SOLVE

$$Q = \frac{V}{t} = Av$$

$$v = \frac{V}{At} = \frac{\left(2.5 \times 10^{-3} \text{ L} \times \frac{1 \text{ m}^3}{1000 \text{ L}}\right)}{\pi \left(\frac{0.114 \times 10^{-3} \text{ m}}{2}\right)^2 (0.65 \text{ s})} = \boxed{380 \frac{\text{m}}{\text{s}}}$$

REFLECT

This large speed is a reason why most injections are given very slowly.

Get Help: P'Cast 11.8 – Blood Flow
P'Cast 11.9 – From Capillaries to the Vena Cavae

11.85

SET UP

At one point, the maximum wind speed during Hurricane Katrina was 240 km/hr and the pressure in the eye was 0.877 atm. We can use the Bernoulli equation and the measured wind speed to calculate the theoretical pressure inside the eye. We will assume the pressure is 1 atm and that the air is still outside of the eye.

SOLVE

$$\frac{1}{2}\rho v_1^2 + P_1 + \rho g y_1 = \frac{1}{2}\rho v_2^2 + P_2 + \rho g y_2$$

$$0 + P_{\text{atm}} + 0 = \frac{1}{2}\rho v_{\text{eye}}^2 + P_{\text{eye}} + 0$$

$$P_{\text{eye}} = P_{\text{atm}} - \frac{1}{2}\rho v_{\text{eye}}^2 = (1.01 \times 10^5 \text{ Pa}) - \frac{1}{2}\left(1.3\frac{\text{kg}}{\text{m}^3}\right)\left(240\frac{\text{km}}{\text{hr}} \times \frac{1000 \text{ m}}{1 \text{ km}} \times \frac{1 \text{ hr}}{3600 \text{ s}}\right)$$

$$= 98{,}111 \text{ Pa} \times \frac{1 \text{ atm}}{1.01 \times 10^5 \text{ Pa}} = \boxed{0.971 \text{ atm}}$$

This is larger than the measured value of 0.877 atm. The pressure outside of the eye is most likely not 1 atm.

REFLECT

The wind is made to go in a circle by the low pressure in the still center, so most of the pressure drop is due to a lower baseline pressure (that is, less than 1 atm), not the Bernoulli effect.

11.89

SET UP

Blood takes about 1.50 s to pass through a capillary ($L = 2.00 \times 10^{-3}$ m) in the human circulatory system. The capillary has a radius of $r = 2.50 \times 10^{-6}$ m. The pressure drop across the capillary is $\Delta P = 2.60 \times 10^3$ Pa. We can use Poiseuille's law in order to calculate the viscosity of the blood. The volumetric flow rate is equal to the volume of the blood in the capillary, which is essentially equal to the volume of the capillary, divided by the time it takes the blood to pass through the capillary. For simplicity, we'll assume the capillary is a cylinder.

SOLVE

$$Q = \frac{\Delta P \pi r^4}{8 \eta L}$$

$$\eta = \frac{\Delta P \pi r^4}{8LQ} = \frac{\Delta P \pi r^4}{8L\left(\dfrac{V}{t}\right)} = \frac{\Delta P \pi r^4 t}{8L(\pi r^2 L)} = \frac{\Delta P r^2 t}{8L^2} = \frac{\Delta P \left(\dfrac{D}{2}\right)^2 t}{8L^2}$$

$$= \frac{\Delta P D^2 t}{32L^2} = \frac{(2.60 \times 10^3 \text{ Pa})(5.00 \times 10^{-6} \text{ m})^2(1.50 \text{ s})}{32(2.00 \times 10^{-3} \text{ m})^2} = \boxed{7.62 \times 10^{-4} \frac{\text{N} \cdot \text{s}}{\text{m}^2}}$$

REFLECT

This is a little less than the value listed in Table 11-2 for the viscosity of blood at body temperature $\left(\sim 3 \times 10^{-4} \dfrac{\text{N} \cdot \text{s}}{\text{m}^2}\right)$.

11.95

SET UP

A normal blood pressure is reported as 120/80, where the top number is the systolic pressure and the bottom number is the diastolic pressure. Both of these values are given in mmHg. We can rewrite these pressures in terms of other pressure units by applying the following conversion factors: 760 mmHg = 1.01×10^5 Pa = 1 atm = 14.7 psi. Blood pressure is an example of a gauge pressure, as 120 mmHg and 80 mmHg refer to the pressures *above* atmospheric pressure.

SOLVE

Part a)

Systolic:

$$120 \text{ mmHg} \times \frac{1.01 \times 10^5 \text{ Pa}}{760 \text{ mmHg}} = 1.59 \times 10^4 \text{ Pa}$$

Diastolic:

$$80 \text{ mmHg} \times \frac{1.01 \times 10^5 \text{ Pa}}{760 \text{ mmHg}} = 1.06 \times 10^4 \text{ Pa}$$

Blood pressure:

$$\frac{1.59 \times 10^4 \text{ Pa}}{1.06 \times 10^4 \text{ Pa}}$$

Part b)

Systolic:

$$120 \text{ mmHg} \times \frac{1 \text{ atm}}{760 \text{ mmHg}} = 0.158 \text{ atm}$$

Diastolic:

$$80 \text{ mmHg} \times \frac{1 \text{ atm}}{760 \text{ mmHg}} = 0.105 \text{ atm}$$

Blood pressure:

$$\frac{0.158 \text{ atm}}{0.105 \text{ atm}}$$

Part c)

Systolic:

$$120 \text{ mmHg} \times \frac{14.7 \text{ psi}}{760 \text{ mmHg}} = 2.32 \text{ psi}$$

Diastolic:

$$80 \text{ mmHg} \times \frac{14.7 \text{ psi}}{760 \text{ mmHg}} = 1.55 \text{ psi}$$

Blood pressure:

$$\frac{2.32 \text{ psi}}{1.55 \text{ psi}}$$

Part d) Blood pressure is reported as a gauge pressure. When you're cut, blood comes out of your arteries; the air doesn't rush in.

REFLECT

If the blood pressures reported were absolute pressures, these would be much smaller than atmospheric pressure. In that case, the pressure difference between the vessels and the outside air would compress all of our blood vessels shut, which luckily does not happen.

11.99

SET UP

A syringe has an inner diameter of 0.6×10^{-3} m. A nurse uses this syringe to inject fluid into a patient's artery where the blood pressure is 140/100. The minimum force the nurse needs to apply to the syringe during the injection is equal to the diastolic blood pressure multiplied by the cross-sectional area of the syringe.

SOLVE

$$F = PA = \left(100 \text{ mmHg} \times \frac{1.01 \times 10^5 \text{ Pa}}{760 \text{ mmHg}}\right)(\pi)\left(\frac{0.6 \times 10^{-3} \text{ m}}{2}\right)^2 = \boxed{0.0038 \text{ N}}$$

REFLECT

The minimum force required corresponds to the minimum pressure in the artery, which is the diastolic pressure.

11.105

SET UP

A faucet is turned on, and the stream of water falls into the sink. The radius and the speed of the stream as it first leaves the faucet are r_0 and v_0, respectively. We can use the Bernoulli equation to find an expression for the speed of the stream v as a function of y, $v(y) = v_1$, where y is the distance the faucet is above the point in the stream. Once we have the speed at an arbitrary point, we can use the continuity equation to find the radius of the stream for an arbitrary point, $r(y) = r_1$. The rate at which the radius changes in terms of y is equal to the time derivative of r_1. Since y is a function of t, we will need to use the chain rule. After we find the expressions for r_1 and dr_1/dt, we can plug in $v_0 = 1.5$ m/s, $r_0 = 0.01$ m, to find the radius and its rate of change at $y = 0.10$ m.

Figure 11-2 Problem 105

SOLVE

Part a)

Bernoulli's equation:

$$P_0 + \frac{1}{2}\rho v_0^2 + \rho g y_0 = P_1 + \frac{1}{2}\rho v_1^2 + \rho g y_1$$

$$P_{\text{atm}} + \frac{1}{2}\rho v_0^2 + \rho g y_0 = P_{\text{atm}} + \frac{1}{2}\rho v_1^2 + \rho g y_1$$

$$v_1 = \sqrt{v_0^2 + 2g(y_0 - y_1)} = \sqrt{v_0^2 + 2gy}$$

Continuity equation:

$$A_0 v_0 = A_1 v_1$$

$$(\pi r_0^2)v_0 = (\pi r_1^2)(\sqrt{v_0^2 + 2gy})$$

$$r_1 = \sqrt{\frac{v_0 r_0^2}{\sqrt{v_0^2 + 2gy}}} = \left(\frac{v_0 r_0^2}{\sqrt{v_0^2 + 2gy}}\right)^{\frac{1}{2}}$$

Speed as a function of y:

$$\frac{dr_1}{dt}(y) = \frac{dr_1}{dy}\frac{dy}{dt} = \frac{dr_1}{dy}(v_1) = \frac{d}{dy}\left[\left(\frac{v_0 r_0^2}{\sqrt{v_0^2 + 2gy}}\right)^{\frac{1}{2}}\right]v_1 = \frac{d}{dy}\left[\left(\frac{1}{\sqrt{v_0^2 + 2gy}}\right)^{\frac{1}{2}}\right]v_1\sqrt{v_0 r_0^2}$$

$$= v_1\sqrt{v_0 r_0^2}\frac{d}{dy}[(v_0^2 + 2gy)^{-\frac{1}{4}}] = v_1\sqrt{v_0 r_0^2}\left[\left(-\frac{1}{4}\right)(v_0^2 + 2gy)^{-\frac{5}{4}}(2g)\right]$$

$$= (v_0^2 + 2gy)^{\frac{1}{2}}\sqrt{v_0 r_0^2}\left[\left(-\frac{g}{2}\right)(v_0^2 + 2gy)^{-\frac{5}{4}}\right] = \boxed{\frac{-g\sqrt{v_0 r_0^2}}{2(v_0^2 + 2gy)^{\frac{3}{4}}}}$$

Part b)

Radius:

$$r_1 = \left(\frac{v_0 r_0^2}{\sqrt{v_0^2 + 2gy}}\right)^{\frac{1}{2}} = \left(\frac{\left(1.5\frac{m}{s}\right)(0.01\text{ m})^2}{\sqrt{\left(1.5\frac{m}{s}\right)^2 + 2\left(9.8\frac{m}{s^2}\right)(0.1\text{ m})}}\right)^{\frac{1}{2}} = \boxed{0.00855\text{ m} = 8.55\text{ mm}}$$

Rate of change of the radius:

$$\frac{dr_1}{dt} = \frac{-g\sqrt{v_0 r_0^2}}{2(v_0^2 + 2gy)^{\frac{3}{4}}} = \frac{-\left(9.8\frac{m}{s^2}\right)\sqrt{\left(1.5\frac{m}{s}\right)(0.01\text{ m})^2}}{2\left(\left(1.5\frac{m}{s}\right)^2 + 2\left(9.8\frac{m}{s^2}\right)(0.1\text{ m})\right)^{\frac{3}{4}}} = \boxed{-0.020\frac{m}{s}}$$

REFLECT

The minus sign in the rate of change of the radius means the water stream is getting narrower the farther it is from the faucet. You can observe this effect in your own kitchen!

Chapter 12
Oscillations

Conceptual Questions

12.1 Oscillatory motion not only includes simple harmonic motion, but it also includes circular motion, decaying oscillations, and oscillations that have shapes other than pure sine waves.

12.5 Breathing rate (breaths per minute) is a frequency. The period is its reciprocal.

12.11 The force of gravity depends on elevation and determines the period of the pendulum. Measuring the period and length of the pendulum will yield enough information to estimate the strength of gravity; doing so very precisely would enable useful comparison to established values.

12.15 The frequency of the driving force is just how often the applied force repeats. That depends on things outside the oscillator. The natural frequency of the oscillator is the frequency at which it oscillates most readily or the frequency at which it will oscillate if displaced from equilibrium and released.

Multiple-Choice Questions

12.17 B $(x = 0)$. The kinetic energy of the block is a maximum when the spring is at its equilibrium position.

12.21 A (The period will increase).

$$T_1 = \frac{2\pi}{\omega_1} = 2\pi\sqrt{\frac{m_1}{k}}$$

$$T_2 = \frac{2\pi}{\omega_2} = 2\pi\sqrt{\frac{m_2}{k}} = 2\pi\sqrt{\frac{(2m_1)}{k}} = \sqrt{2}\left(2\pi\sqrt{\frac{m_1}{k}}\right) = \sqrt{2}T_1$$

Get Help: P'Cast 12.2 – Four Times the Mass

Estimation Questions

12.27 Swings are typically around 2 m in length, so we get a period of around 3 s.

Get Help: P'Cast 12.2 – Four Times the Mass

12.33 A 0.01-g fly clinging to the bottom of a pendulum that has a disk bob that is 1 kg, has a radius of 10 cm, and is centered at 1 m from the axis would produce a frequency reduction of around 1 part in 4×10^7.

12.39 On average, the period looks to be about 4 years.

Get Help: P'Cast 12.2 – Four Times the Mass

Problems

12.43

SET UP

An object of mass M is attached to a spring with spring constant k and hung vertically. When the mass is initially hung on the spring, the spring will stretch, and the system will eventually come to rest at a new equilibrium position, which we'll call y_0. When the mass is slightly displaced, it will oscillate about this new equilibrium position. If we displace the mass downward a distance $(y - y_0)$, the forces acting on it are the spring pulling it back up and gravity pulling it down. We can rearrange Newton's second law for the mass and compare it with the standard differential equation for simple harmonic motion to find an expression for the motion of the mass as a function of time.

SOLVE

Newton's second law:

$$\sum F_y = Mg - k(y - y_0) = Ma = M\frac{d^2y}{dt^2}$$

$$\frac{d^2y}{dt^2} + \frac{k}{M}(y - y_0) - g = 0$$

$$\frac{d^2y}{dt^2} + \frac{k}{M}\left(y - y_0 - \frac{Mg}{k}\right) = 0$$

Change of variables; let $u = y - y_0 - \dfrac{Mg}{k}$:

$$\frac{d^2u}{dt^2} + \frac{k}{M}u = 0$$

$$\frac{d^2u}{dt^2} + \omega^2 u = 0$$

$$u = A\cos(\omega t) = y - y_0 - \frac{Mg}{k}$$

$$y = A\cos(\omega t) + \left(y_0 + \frac{Mg}{k}\right)$$

REFLECT

The force of gravity effectively changes the equilibrium position of a spring–mass system undergoing simple harmonic motion in the vertical direction. Gravity has no effect on the frequency of the oscillation.

12.49

SET UP

The mathematical description of the motion of a particular simple harmonic oscillator is
$x(t) = (0.15 \text{ m})\cos\left(\pi t + \dfrac{\pi}{3}\right)$. The velocity of the oscillator is equal to the first derivative
of the position with respect to time, and the acceleration of the oscillator is equal to the first
derivative of the velocity with respect to time. We will assume that all of the values are given
in SI units.

SOLVE

$$x(t) = (0.15 \text{ m})\cos\left(\pi t + \frac{\pi}{3}\right)$$

Part a)

$$v(t) = \frac{dx}{dt} = \frac{d}{dt}\left[(0.15 \text{ m})\cos\left(\pi t + \frac{\pi}{3}\right)\right] = (0.15 \text{ m})\left[-\pi\sin\left(\pi t + \frac{\pi}{3}\right)\right]$$

$$v(t = 1 \text{ s}) = (0.15 \text{ m})\left[-\pi\sin\left(\pi(1 \text{ s}) + \frac{\pi}{3}\right)\right] = \boxed{0.41\frac{\text{m}}{\text{s}}}$$

Part b)

$$a(t) = \frac{dv}{dt} = \frac{d}{dt}\left[-(0.15 \text{ m})\pi\sin\left(\pi t + \frac{\pi}{3}\right)\right] = -(0.15 \text{ m})\pi^2\cos\left(\pi t + \frac{\pi}{3}\right)$$

$$a(t = 2 \text{ s}) = -(0.15 \text{ m})\pi^2\cos\left(\pi(2 \text{ s}) + \frac{\pi}{3}\right) = \boxed{-0.74\frac{\text{m}}{\text{s}^2}}$$

REFLECT

The acceleration of the oscillator is also equal to the second derivative of the position with
respect to time. Since we already calculated the velocity, it's easier to differentiate this function
once than go back to the position and differentiate it twice.

12.53

SET UP

An object undergoing simple harmonic motion has a frequency of $f = 15$ Hz. The period of
the object's motion is equal to the reciprocal of its frequency. Multiplying the frequency of the
motion by 2 min will give us the number of cycles the object undergoes in that time period.

SOLVE

Part a)

$$T = \frac{1}{f} = \frac{1}{15 \text{ Hz}} = \boxed{0.067 \text{ s}}$$

Part b)

$$N_{cycles} = 2 \text{ min} \times \frac{60 \text{ s}}{1 \text{ min}} \times \frac{15 \text{ cycles}}{1 \text{ s}} = \boxed{1800 \text{ cycles}}$$

REFLECT

A frequency of 15 Hz is reasonably fast for a mass on the end of a spring.

Get Help: P'Cast 12.2 – Four Times the Mass

12.57

SET UP

Two springs with spring constants k_1 and k_2, respectively, are attached end-to-end to a box of mass M. We can find the effective spring constant of this setup by comparing the two-spring setup with an equivalent one-spring setup. If the box is moved a distance Δx, each spring will stretch a different amount (Δx_1 and Δx_2) but their sum must be Δx. In the figure, we see that only spring 2 is attached to the mass, which means only the force due to spring 2 acts on the mass. Newton's third law tells us that the magnitude of the force of spring 1 on spring 2 must equal the magnitude of the force of spring 2 on spring 1: $k_1\Delta x_1 = k_2\Delta x_2$; we can use this fact to eliminate Δx_1 from our equation. The period of the motion of the box is equal to

$$T = 2\pi\sqrt{\frac{M}{k_{eff}}}.$$

Figure 12-1 Problem 57

SOLVE

$$\Delta x = \Delta x_1 + \Delta x_2 = \left(\frac{k_2\Delta x_2}{k_1}\right) + \Delta x_2 = \left(\frac{k_2}{k_1} + 1\right)\Delta x_2$$

$$\Delta x_2 = \frac{\Delta x}{\left(\frac{k_2}{k_1} + 1\right)}$$

Effective spring constant:

$$F_2 = F_{eff}$$

$$k_2\Delta x_2 = k_{eff}\Delta x$$

$$k_2\left(\frac{\Delta x}{\left(\frac{k_2}{k_1} + 1\right)}\right) = k_{eff}\Delta x$$

$$k_{eff} = \frac{k_2}{\left(\frac{k_2}{k_1} + 1\right)} = \frac{k_2}{\left(\frac{k_2 + k_1}{k_1}\right)} = \frac{k_1 k_2}{(k_2 + k_1)}$$

Period:

$$T = \frac{2\pi}{\omega} = 2\pi\sqrt{\frac{M}{k_{\text{eff}}}} = \boxed{2\pi\sqrt{\frac{M(k_1 + k_2)}{k_1 k_2}} = 2\pi\sqrt{M\left(\frac{1}{k_1} + \frac{1}{k_2}\right)}}$$

REFLECT

The springs in this configuration are described as being in "series." The effective spring constant in this case is *smaller* than the individual spring constants.

Get Help: P'Cast 12.2 – Four Times the Mass

12.63

SET UP

The mathematical description of an object attached to a spring undergoing simple harmonic motion on a horizontal frictionless surface is $x(t) = A\cos(\omega t + \phi)$. The potential energy of the object is equal to the potential energy stored in the spring, $U = \frac{1}{2}kx^2$. The kinetic energy of the object is $K = \frac{1}{2}mv^2 = \frac{1}{2}m\left(\frac{dx}{dt}\right)^2$. The total energy of a simple harmonic oscillator is given by $E = \frac{1}{2}mv^2 + \frac{1}{2}kx^2$. We can use our answers from parts (a) and (b) and do some algebra in order to show explicitly that E is a constant. We will need to invoke the Pythagorean identity: $\sin^2(\omega t + \phi) + \cos^2(\omega t + \phi) = 1$.

SOLVE

Part a)

$$U = \frac{1}{2}kx^2 = \frac{1}{2}k[A\cos(\omega t + \phi)]^2 = \boxed{\frac{kA^2}{2}\cos^2(\omega t + \phi)}$$

Part b)

$$K = \frac{1}{2}mv^2 = \frac{1}{2}m\left(\frac{dx}{dt}\right)^2 = \frac{1}{2}m\left(\frac{d}{dt}[A\cos(\omega t + \phi)]\right)^2 = \frac{1}{2}m(-\omega A\sin(\omega t + \phi))^2$$

$$= \boxed{\frac{m\omega^2 A^2}{2}\sin^2(\omega t + \phi)}$$

Part c)

$$E = \frac{1}{2}kx^2 + \frac{1}{2}mv^2 = \frac{kA^2}{2}\cos^2(\omega t + \phi) + \frac{m\omega^2 A^2}{2}\sin^2(\omega t + \phi)$$

$$= \frac{kA^2}{2}\cos^2(\omega t + \phi) + \frac{m\left(\frac{k}{m}\right)A^2}{2}\sin^2(\omega t + \phi)$$

$$= \frac{kA^2}{2}\sin^2(\omega t + \phi) + \frac{kA^2}{2}\cos^2(\omega t + \phi)$$

$$= \frac{kA^2}{2}[\sin^2(\omega t + \phi) + \cos^2(\omega t + \phi)] = \boxed{\frac{1}{2}kA^2}$$

This is constant for a given k and A.

REFLECT

Because the object is sliding along a horizontal, frictionless surface we can ignore its gravitational potential energy.

12.65

SET UP

A 0.250-kg object attached to a spring oscillates on a frictionless horizontal table with a frequency of $f = 4$ Hz and an amplitude of $A = 0.20$ m. Although we were not given it explicitly, we can rewrite the spring constant in terms of the frequency and mass. The maximum potential energy of the system occurs when the kinetic energy is equal to zero, which means the maximum potential energy is equal to the total energy of the harmonic oscillator. We can then divide this quantity by two and solve for x to solve for the displacement of the object when the potential energy is one-half of the maximum. Finally, we can plug $x = 0.10$ m into the elastic potential energy to find the potential energy of the object at that displacement.

SOLVE
Part a)

$$U_{max} = \frac{1}{2}kA^2 = \frac{1}{2}\omega^2 mA^2 = \frac{1}{2}(2\pi f)^2 mA^2 = 2\pi^2 mf^2 A^2$$

$$= 2\pi^2(0.250 \text{ kg})(4 \text{ Hz})^2(0.20 \text{ m})^2 = \boxed{3.2 \text{ J}}$$

Part b)

$$U_{half\ max} = \frac{1}{2}kx^2 = \frac{1}{2}\left(\frac{1}{2}kA^2\right)$$

$$x = \frac{A}{\sqrt{2}} = \frac{20 \text{ cm}}{\sqrt{2}} = \boxed{14 \text{ cm}}$$

Part c)

$$U(x) = \frac{1}{2}kx^2 = \frac{1}{2}\omega^2 mx^2 = \frac{1}{2}(2\pi f)^2 mx^2 = 2\pi^2 mf^2 x^2$$

$$U(x = 0.10 \text{ m}) = 2\pi^2(0.250 \text{ kg})(4 \text{ Hz})^2(0.10 \text{ m})^2 = \boxed{0.79 \text{ J}}$$

REFLECT

The displacement where the potential energy is one-half of the maximum value does *not* occur at one-half the amplitude.

12.67

SET UP

The equation of motion for a given simple harmonic oscillator is $x(t) = A\sin(\omega t)$. The velocity of this oscillator is equal to the first derivative of $x(t)$ with respect to time. The potential energy of a simple harmonic oscillator is given by $U(t) = \frac{1}{2}kx^2$ and the kinetic energy is given by $K(t) = \frac{1}{2}mv^2$. The plots of $U(t)$ and $K(t)$ should look like $\sin^2(\omega t)$ and $\cos^2(\omega t)$, respectively.

SOLVE

Part a)

$$U(t) = \frac{1}{2}kx^2 = \frac{1}{2}k[A\sin(\omega t)]^2 = \frac{1}{2}kA^2\sin^2(\omega t)$$

For simplicity, the vertical axis of the following plot is $\dfrac{U}{\left(\frac{1}{2}kA^2\right)}$, and the horizontal axis is ωt:

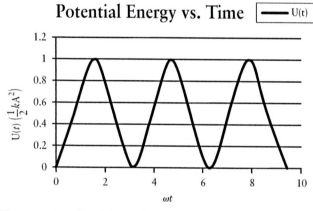

Figure 12-2 Problem 67

Part b)

$$v(t) = \frac{dx}{dt} = \frac{d}{dt}[A\sin(\omega t)] = A\omega\cos(\omega t)$$

Part c)

$$K(t) = \frac{1}{2}mv^2 = \frac{1}{2}mA^2\omega^2\cos^2(\omega t) = \frac{1}{2}kA^2\cos^2(\omega t)$$

For simplicity, the vertical axis of the following plot is $\dfrac{U}{\left(\frac{1}{2}kA^2\right)}, \dfrac{K}{\left(\frac{1}{2}kA^2\right)}$ and the horizontal axis is ωt:

Potential Energy, Kinetic Energy vs. Time

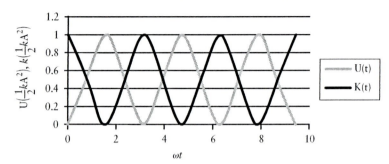

Figure 12-3 Problem 67

REFLECT

Although U and K oscillate in time, their sum is constant for all time.

12.71

SET UP

A simple pendulum completes 14 cycles in 25 s. We can use the definition of the period of a simple pendulum $T = 2\pi\sqrt{\dfrac{L}{g}}$ and solve for its length.

SOLVE

$$T = 2\pi\sqrt{\frac{L}{g}}$$

$$L = g\left(\frac{T}{2\pi}\right)^2 = \left(9.8\frac{m}{s^2}\right)\left(\frac{\left(\dfrac{25\ s}{14\ cycles}\right)}{2\pi}\right)^2 = \boxed{0.792\ m}$$

REFLECT

The period of a pendulum is the time it takes to make one full oscillation.

12.75

SET UP

A simple pendulum has a length of $L = 0.50$ m and oscillates with an amplitude of 8 degrees. From these data we can calculate the period of the motion and write down the equation of motion for the pendulum. The time it takes the pendulum to swing from $+8$ degrees to -8 degrees is half of a period. To find the time interval between $+4$ degrees and -4 degrees, we will need to find the times at which the pendulum is at each of these angular positions; we can do this by solving the equation of motion.

SOLVE

Period:

$$T = 2\pi\sqrt{\frac{L}{g}} = 2\pi\sqrt{\frac{0.50\ m}{\left(9.8\dfrac{m}{s^2}\right)}} = 1.42\ s$$

Equation of motion:

$$\theta(t) = \theta_0 \cos(\omega t) = \theta_0 \cos\left(\frac{2\pi t}{T}\right) = (8°)\cos\left(\left(4.42\frac{\text{rad}}{\text{s}}\right)t\right)$$

Time interval between $+8°$ and $-8°$:

$$\frac{T}{2} = \boxed{0.71 \text{ s}}$$

Time interval between $+4°$ and $-4°$:

$$4° = (8°)\cos\left(\left(4.42\frac{\text{rad}}{\text{s}}\right)t_1\right)$$

$$t_1 = \frac{\arccos\left(\dfrac{4°}{8°}\right)}{\left(4.42\dfrac{\text{rad}}{\text{s}}\right)} = 0.237 \text{ s}$$

$$-4° = (8°)\cos\left(\left(4.42\frac{\text{rad}}{\text{s}}\right)t_2\right)$$

$$t_2 = \frac{\arccos\left(\dfrac{-4°}{8°}\right)}{\left(4.42\dfrac{\text{rad}}{\text{s}}\right)} = 0.474 \text{ s}$$

$$\Delta t = t_2 - t_1 = (0.474 \text{ s}) - (0.237 \text{ s}) = \boxed{0.237 \text{ s}}$$

REFLECT

The time interval between $+4°$ and $-4°$ is one-third of the period.

12.79

SET UP

A solid sphere of mass M and radius $R = 0.05$ m is suspended from an eyelet attached to its surface. The sphere is displaced slightly from equilibrium and set into simple harmonic motion. The period of a physical pendulum is related to the moment of inertia of the pendulum. Since the center of mass of the sphere is located a distance $h = R$ from the pivot point, we will need to use the parallel-axis theorem. As a reminder, the moment of inertia of a sphere is $\frac{2}{5}MR^2$.

Figure 12-4 Problem 79

SOLVE

$$T = 2\pi\sqrt{\frac{I}{mgh}} = 2\pi\sqrt{\frac{I_{\text{sphere}} + I_{\text{from axis}}}{mgh}} = 2\pi\sqrt{\frac{\left(\frac{2}{5}MR^2\right) + MR^2}{MgR}} = 2\pi\sqrt{\frac{7R}{5g}}$$

$$= 2\pi\sqrt{\frac{7(0.05 \text{ m})}{5\left(9.8\frac{\text{m}}{\text{s}^2}\right)}} = \boxed{0.53 \text{ s}}$$

REFLECT

The density of the material is irrelevant since the mass cancels out in our expression for the period of the oscillation.

12.83

SET UP

A pendulum bob of mass $m = 0.110$ kg swings at the end of a wire of length $L = 15.0$ m. The damping coefficient of the pendulum is $b = 0.010$ kg/s. We can use the definition of the lightly damped and undamped oscillation frequencies to calculate the period of this damped harmonic oscillator. The amplitude of a damped harmonic oscillator is a function of time: $A(t) = A_0 e^{-\frac{b}{2m}t}$. We can use this relationship to find the amplitude of the pendulum after a time interval of three periods.

SOLVE

Period:

$$T = \frac{2\pi}{\omega_1} = \frac{2\pi}{\sqrt{\omega_0^2 - \frac{b^2}{4m^2}}} = \frac{2\pi}{\sqrt{\frac{g}{L} - \frac{b^2}{4m^2}}} = \frac{2\pi}{\sqrt{\frac{\left(9.8\frac{\text{m}}{\text{s}^2}\right)}{(15.0 \text{ m})} - \frac{\left(0.010\frac{\text{kg}}{\text{s}}\right)^2}{4(0.110 \text{ kg})^2}}} = 7.79 \text{ s}$$

Amplitude:

$$A(t) = A_0 e^{-\frac{b}{2m}t}$$

$$A(3T) = A_0 e^{-\frac{b}{2m}(3T)} = (1.5 \text{ m})e^{-\frac{(0.010\frac{\text{kg}}{\text{s}})(3)(7.79 \text{ s})}{2(0.110 \text{ kg})}} = \boxed{0.52 \text{ m}}$$

REFLECT

The amplitude of the pendulum exponentially decays with time and comes to rest after an infinitely long period of time.

Get Help: Picture It: Damped Oscillations

12.87

SET UP

An oscillating system has a natural frequency of $\omega_0 = 50$ rad/s and a damping coefficient of $b = 2.0$ kg/s. The system is driven by an oscillating force, $F(t) = (100 \text{ N})\cos\left(\left(50\frac{\text{rad}}{\text{s}}\right)t\right)$.

From this expression, we can see that the frequency of the applied force is equal to the natural frequency of the system. The amplitude of a damped, driven oscillator is

$$A = \frac{F_0}{m\sqrt{(\omega_0^2 - \omega^2)^2 + \frac{b^2\omega^2}{m^2}}}.$$

SOLVE

$$A = \frac{F_0}{m\sqrt{(\omega_0^2 - \omega^2)^2 + \left(\frac{b\omega}{m}\right)^2}} = \frac{F_0}{m\sqrt{(\omega_0^2 - \omega_0^2)^2 + \left(\frac{b\omega_0}{m}\right)^2}} = \frac{F_0}{b\omega_0}$$

$$= \frac{100 \text{ N}}{\left(2.0\frac{\text{kg}}{\text{s}}\right)\left(50\frac{\text{rad}}{\text{s}}\right)} = \boxed{1.0 \text{ m}}$$

REFLECT

This is the maximum amplitude of the system's motion because the applied force oscillates at the natural frequency of the system.

12.91

SET UP

The acceleration of an object ($m = 0.025$ kg) that exhibits simple harmonic motion is given by $a(t) = \left(10\frac{\text{m}}{\text{s}^2}\right)\cos\left(\pi t + \frac{\pi}{2}\right)$. We can determine the expression for the object's velocity $v(t)$ by integrating $a(t)$ with respect to time. We are told that the object's speed is a maximum at $t = 0$. Once we have the algebraic expression for $v(t)$, we can calculate $v(t = 2$ s$)$.

SOLVE

$$v(t) = \int a(t)dt = \int 10\cos\left(\pi t + \frac{\pi}{2}\right)dt = \frac{10}{\pi}\sin\left(\pi t + \frac{\pi}{2}\right) + C \text{ (SI units)}$$

The maximum speed of an object undergoing simple harmonic motion is equal to its amplitude, which is $\frac{10}{\pi}\frac{\text{m}}{\text{s}}$ in this case. Therefore, $C = 0$, so

$$v(t) = \frac{10}{\pi}\sin\left(\pi t + \frac{\pi}{2}\right) \text{ (SI units)}$$

Velocity at $t = 2$ s:

$$v(2) = \frac{10}{\pi}\sin\left(2\pi + \frac{\pi}{2}\right) = \boxed{\frac{10}{\pi}\frac{\text{m}}{\text{s}} = 3.18\frac{\text{m}}{\text{s}}}$$

REFLECT

The general equation of the acceleration in terms of the period of the oscillation is $a(t) = a_0\cos\left(\frac{2\pi t}{T} + \phi\right)$. The period of the oscillation in this problem is $T = 2$ s, which means the velocity at $t = 0$ s will be equal to the velocity at $t = 2$ s.

12.95

SET UP

A damped oscillator shows a reduction of 30% in amplitude after two periods. We can use the amplitude of a damped oscillator as a function of time, $A(t) = A_0 e^{-\frac{b}{2m}t}$, to find the reduction in amplitude and the mechanical energy loss after one period.

SOLVE

Amplitude after one period:

$$\frac{A_{2T}}{A_0} = e^{-\frac{b}{2m}(2T)} = \left(e^{-\frac{b}{2m}(T)}\right)^2 = \left(\frac{A_T}{A_0}\right)^2 = 0.7$$

Percent loss in mechanical energy per cycle:

$$\left(\frac{E_T - E_0}{E_0}\right) = \frac{\left(\frac{1}{2}kA_T^2\right) - \left(\frac{1}{2}kA_0^2\right)}{\left(\frac{1}{2}kA_0^2\right)} = \left(\frac{A_T}{A_0}\right)^2 - 1 = (0.837)^2 - 1 = -0.3$$

There is a $\boxed{30\% \text{ loss}}$ of mechanical energy per cycle.

REFLECT

The change in mechanical energy of the system is equal to the work done by all nonconservative forces on the system.

12.101

SET UP

A spinning golf ball of radius R can be suspended in a stream of high-velocity air. The ball is in equilibrium at the vertical center of the stream, which we'll call $y = 0$. The forces acting on the ball are the net pressure acting up and the force of gravity pointing down; at equilibrium, these forces balance. There is a difference in pressure due to the flowing air, which we can relate to the speed of the air above and below the ball using Bernoulli's equation. The ball is spinning at a speed v_{rot}. This will cause the air to be dragged along with the spinning ball, which affects the speed of the air above and below the ball. Also, the speed of the air drops off linearly from the center of the stream, that is, $v(y) = v_0 - b|y|$, where v_0 is the speed of the air in the center and b is a constant. If the ball is displaced a small amount Δy below equilibrium, the ball will no longer be in equilibrium, and there will be a net force upward on the ball since the ball is located in a different part of the air stream now. In order to show that the ball undergoes simple harmonic motion in this case, we need to show that the net force on the ball has the form of Hooke's law: The net force is proportional to the distance the ball is displaced from equilibrium.

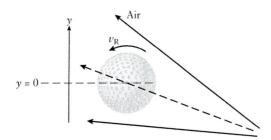

Figure 12-5 Problem 101

SOLVE

Free-body diagram of the spinning golf ball:

Figure 12-6 Problem 101

Magnitude of $\vec{F}_{\text{up}}$:

$$F_{\text{up}} = (P_{\text{below}} - P_{\text{above}})A$$

Using Bernoulli's equation, $P_{\text{above}} + \frac{1}{2}\rho v_{\text{above}}^2 = P_{\text{below}} + \frac{1}{2}\rho v_{\text{below}}^2$:

$$F_{\text{up}} = (P_{\text{below}} - P_{\text{above}})A = \left(\frac{1}{2}\rho v_{\text{above}}^2 - \frac{1}{2}\rho v_{\text{below}}^2\right)A = \frac{1}{2}\rho A(v_{\text{above}}^2 - v_{\text{below}}^2)$$

Newton's second law at equilibrium:

$$\sum F_y = F_{\text{up}} - F_g = \frac{1}{2}\rho A(v_{\text{above}}^2 - v_{\text{below}}^2) - mg = ma_y = 0$$

$$\frac{1}{2}\rho A(v_{\text{above}}^2 - v_{\text{below}}^2) = mg$$

But the speeds above and below the ball are $v_{\text{above}} = v(+R) + v_{\text{rot}} = (v_0 - bR) + v_{\text{rot}}$ and $v_{\text{below}} = v(-R) - v_{\text{rot}} = (v_0 - b|-R|) - v_{\text{rot}} = (v_0 - bR) - v_{\text{rot}}$. Plugging these into our expression from Newton's second law:

$$\frac{1}{2}\rho A(v_{\text{above}}^2 - v_{\text{below}}^2) = \frac{1}{2}\rho A((v_0 - bR + v_{\text{rot}})^2 - (v_0 - bR - v_{\text{rot}})^2) = mg$$

$$\frac{1}{2}\rho A(v_0^2 - 2v_0bR - 2v_{\text{rot}}bR + 2v_0v_{\text{rot}} + b^2R^2 + v_{\text{rot}}^2 - (v_0^2 - 2v_0bR + 2v_{\text{rot}}bR - 2v_0v_{\text{rot}} + b^2R^2 + v_{\text{rot}}^2)) = mg$$

$$\frac{1}{2}\rho A(4v_0v_{\text{rot}} - 4v_{\text{rot}}bR) = mg$$

If the ball is displaced downward an amount Δy, the upward force no longer equals mg, which results in a vertical net force upward:

$$\sum F_y = F_{\text{up}} - mg = \frac{1}{2}\rho A(v_{\text{above}}^2 - v_{\text{below}}^2) - mg$$

But the speeds above and below the ball are $v_{\text{above}} = v(R - \Delta y) + v_{\text{rot}} = (v_0 - b(R - \Delta y)) + v_{\text{rot}}$ and $v_{\text{below}} = v(-R - \Delta y) - v_{\text{rot}} = (v_0 - b|-R - \Delta y|) - v_{\text{rot}} = (v_0 - b(R + \Delta y)) - v_{\text{rot}}$. Plugging these into our expression from Newton's second law:

$$\frac{1}{2}\rho A((v_0 - bR + b(\Delta y) + v_{\text{rot}})^2 - (v_0 - bR - b(\Delta y) - v_{\text{rot}})^2) - mg$$

$$= \frac{1}{2}\rho A(4v_0 b(\Delta y) + 4v_0 v_{\text{rot}} - 4v_{\text{rot}}bR - 4b^2 R(\Delta y)) - mg$$

$$= \frac{1}{2}\rho A(4v_0 b(\Delta y) + 4v_0 v_{\text{rot}} - 4v_{\text{rot}}bR - 4b^2 R(\Delta y)) - \left[\frac{1}{2}\rho A(4v_0 v_{\text{rot}} - 4v_{\text{rot}}bR)\right]$$

$$= \frac{1}{2}\rho A(4v_0 b(\Delta y) - 4b^2 R(\Delta y)) = -2\rho A b(bR - v_0)(\Delta y)$$

This has the form $\sum F_y = -k(\Delta y)$, which is Hooke's law.

REFLECT

As with a mass on a spring, we can interpret the coefficient in front of Δy as a spring constant of sorts and relate it to the frequency of oscillation of the ball.

12.109

SET UP

A string ($L_0 = 0.200$ m) is made by combining 10 parallel strands of hair that are each 1.25×10^{-4} m thick. The Young's modulus of the hair is 4.50×10^9 Pa. We can calculate the effective spring constant k for an individual hair from the Young's modulus and its dimensions. The effective spring constant for the string made up of 10 parallel hairs is $k_{\text{eff}} = 10k$. A 0.175-kg utensil is hung from one end of the string. When it comes to rest, the elastic force of the string pulling up must equal the force of gravity pulling down. We can calculate the amount the string stretches from this force balance. The utensil is pulled down slightly and set into simple harmonic motion. The amount of time it takes the utensil to first return to its initial position is equal to one period, which we can calculate from the mass of the utensil and the effective spring constant of the string.

SOLVE

Part a)

$$F_g = F_{\text{hair}}$$

$$mg = k_{\text{eff}}\Delta L = (10k)\Delta L = 10\left(\frac{AY}{L_0}\right)\Delta L = 10(\pi R^2)\left(\frac{Y}{L_0}\right)\Delta L = 10\pi\left(\frac{D}{2}\right)^2\left(\frac{Y}{L_0}\right)\Delta L$$

$$\Delta L = \frac{2mgL_0}{5\pi D^2 Y} = \frac{2(0.175 \text{ kg})\left(9.8\frac{\text{m}}{\text{s}^2}\right)(0.200 \text{ m})}{5\pi(125 \times 10^{-6} \text{ m})^2(4.50 \times 10^9 \text{ Pa})} = \boxed{6.21 \times 10^{-4} \text{ m} = 621 \text{ }\mu\text{m}}$$

Part b)

$$T = \frac{2\pi}{\omega} = 2\pi\sqrt{\frac{m}{k_{\text{eff}}}} = 2\pi\sqrt{\frac{m}{10k}} = 2\pi\sqrt{\frac{m}{10\left(\frac{AY}{L_0}\right)}} = 2\pi\sqrt{\frac{mL_0}{10(\pi R^2)Y}} = 2\sqrt{\frac{\pi m L_0}{10\left(\frac{D}{2}\right)^2 Y}}$$

$$= 2\sqrt{\frac{2\pi m L_0}{5D^2 Y}} = 2\sqrt{\frac{2\pi(0.175 \text{ kg})(0.200 \text{ m})}{5(125 \times 10^{-6} \text{ m})^2(4.50 \times 10^9 \text{ Pa})}} = \boxed{0.050 \text{ s} = 50.0 \text{ ms}}$$

REFLECT

Spring constants add normally for springs in parallel, while they add in reciprocal for springs in series.

Get Help: P'Cast 12.2 – Four Times the Mass

12.111

SET UP

A baseball of mass m is dropped through a narrow tunnel that was drilled through the center of the Earth. The only force acting on the baseball while inside the Earth is the force due to gravity. According to the Shell Theorem, the gravitational field inside of a sphere is $g(r) = -g_0\left(\dfrac{r}{R}\right)$, where $g_0 = 9.8$ m/s², R is the radius of the Earth $(6.38 \times 10^6$ m$)$, and r is the distance from the center of the Earth. Because the gravitation field is proportional to the distance, we can rearrange Newton's second law such that it resembles the simple harmonic motion differential equation. This allows us to find the angular frequency and, thus, the period for the motion.

SOLVE

Free-body diagram for the ball above the center of the Earth:

Figure 12-7 Problem 111

Free-body diagram for the ball below the center of the Earth:

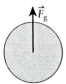

Figure 12-8 Problem 111

Newton's second law for the ball:

$$\sum F_r = F_g = -mg = -mg_0\left(\frac{r}{R}\right) = ma = m\frac{d^2r}{dt^2}$$

$$\frac{d^2r}{dt^2} + \left(\frac{g_0}{R}\right)r = 0$$

$$\omega^2 = \frac{g_0}{R}$$

$$T = \frac{2\pi}{\omega} = 2\pi\sqrt{\frac{R}{g_0}} = 2\pi\sqrt{\frac{(6.38 \times 10^6 \text{ m})}{\left(9.8\frac{\text{m}}{\text{s}^2}\right)}} = \boxed{5070 \text{ s}}$$

REFLECT

It takes about 1 hour and 25 minutes for the ball to return to its initial position.

Chapter 13
Waves

Conceptual Questions

13.3 A very few of them might be the same molecules, especially if he is close by. The air molecules collide many times on the way from place to place. The collisions transmit the wave motion without requiring the individual molecules to go the full distance.

13.7 Part a) The wave speed depends entirely on the medium so wave speed increases as the wave passes from air into water.

Part b) The frequency is determined by the actual vibrations of the source, so the frequency does not change.

Part c) The wavelength depends on changes in the wave speed, so the wavelength increases, too.

13.11 The frequency counts the number of oscillations per time. The angular frequency counts the oscillations per time in units of 2π; one complete oscillation is equivalent to 2π rad. The angular frequency is more natural from a mathematical point of view.

13.17 The time derivative of the transverse position y is how far from equilibrium any one component is. In particular, the velocity is proportional to the amplitude. The wave speed is not. It is how fast any part of the wave (the pattern, not the constituent parts) moves.

13.23 Part a) Intensity is the power per unit area that is emitted by a source of sound. It has nothing to do with the reception of the sound through the process of "hearing." The SI units are W/m^2. Sound level is a mathematical relationship that puts the intensity onto a logarithmic scale. The units are decibels (dB). Loudness is the physiological response to sound waves. Different frequencies with the same intensity will be more sensitively heard by humans. Power is the energy per unit time emitted by the source.

Part b) Intensity depends on the inverse of the distance squared, so the intensity increases as the source of sound moves closer.

Part c) Sound level depends on the log of intensity, so sound level increases as the source of sound moves closer.

Part d) The total power emitted by the source does not change as it moves closer to the observer.

Multiple-Choice Questions

13.29 D (5).

Figure 13-1 Problem 29

13.33 D (The sound intensity drops to 1/9 its original value). The sound intensity is inversely proportional to the square of the distance.

Estimation Questions

13.37 The wavelength of sound with a frequency of 3000 Hz is (343 m/s)/(3000 Hz) = 0.114 m = 11.4 cm. If the voice box is considered an open–closed organ pipe,

$$L = \frac{\lambda}{4} = 2.8 \text{ cm}.$$

13.41 We can measure the pitch of the fundamental compared to a known standard (like the note A 440 Hz) and measure the length of the string. The speed is twice the length divided by the frequency.

Problems

13.47

SET UP

A wave on a string propagates at $v = 22$ m/s. The frequency of the wave is $f = 24$ Hz. We can use the speed of the wave and the frequency to calculate the wavelength. The wavenumber is $k = \dfrac{2\pi}{\lambda}$.

SOLVE

Wavelength:

$$\lambda = \frac{v}{f} = \frac{\left(22\dfrac{\text{m}}{\text{s}}\right)}{24 \text{ Hz}} = \boxed{0.92 \text{ m} = 92 \text{ cm}}$$

Wavenumber:

$$k = \frac{2\pi}{\lambda} = \frac{2\pi}{\left(\dfrac{v}{f}\right)} = \frac{2\pi f}{v} = \frac{2\pi(24 \text{ Hz})}{\left(22\dfrac{\text{m}}{\text{s}}\right)} = \boxed{6.85\frac{\text{rad}}{\text{m}}}$$

REFLECT

Be careful when rounding intermediate values. You should carry along one more significant figure than you are allowed in your final answer throughout your calculations.

13.51

SET UP

A wave on a string is described by $y(x, t) = 0.05\sin(x - 10t)$ (SI units). We can compare this equation to the general form of a transverse wave, $y(x, t) = A\sin(kx - \omega t + \phi)$, along with the definitions of the wavenumber and angular frequency in order to determine the frequency, the wavelength, and the speed of the wave.

SOLVE

Part a)

$$f = \frac{\omega}{2\pi} = \frac{\left(10\frac{\text{rad}}{\text{s}}\right)}{2\pi} = \boxed{1.6 \text{ Hz}}$$

Part b)

$$\lambda = \frac{2\pi}{k} = \frac{2\pi}{1 \text{ m}^{-1}} = \boxed{6.3 \text{ m}}$$

Part c)

$$v = \frac{\omega}{k} = \frac{\left(10\frac{\text{rad}}{\text{s}}\right)}{1 \text{ m}^{-1}} = \boxed{10\frac{\text{m}}{\text{s}}}$$

REFLECT

We could have also used $v = \lambda f$ to find the speed, but it's better to use the values provided rather than intermediate values that you've calculated.

13.55

SET UP

We are given plots of a wave at $t = 0$ and $x = 0$. Since the plots look like sine functions, we will use this in our mathematical description. We can read the amplitude, wavelength, and period directly from the graphs. The wavenumber and angular frequency can be calculated from the wavelength and period. Because the wave starts at $y = 0$, the phase is also equal to 0.

$t = 0$ s:

$x = 0$ m:

Figure 13-2 Problem 55

SOLVE

Amplitude, $A = 0.10$ m

Wavelength, $\lambda = 0.20$ m

Period, $T = 2$ s

Wavenumber:

$$k = \frac{2\pi}{\lambda} = \frac{2\pi}{(0.20 \text{ m})} = 10\pi \text{ m}^{-1}$$

Angular frequency:

$$\omega = \frac{2\pi}{T} = \frac{2\pi}{2 \text{ s}} = \pi\frac{\text{rad}}{\text{s}}$$

Mathematical description:

$$\boxed{y(x, t) = 0.10\sin(10\pi x - \pi t)} \text{ (SI units)}$$

REFLECT

The plot y versus t looks like $-\sin(\omega t)$ because sine is an odd function; that is, $\sin(-\omega t) = -\sin(\omega t)$.

13.57

SET UP

A string has a mass of 5.0×10^{-3} kg and length of 2.2 m. The string is pulled taut with a tension of $T = 78$ N. The propagation speed of a transverse wave on a string is given by $v_p = \sqrt{\dfrac{T}{\mu}}$, where T is the tension and μ is the linear mass density.

SOLVE

$$v_p = \sqrt{\frac{T}{\mu}} = \sqrt{\frac{78 \text{ N}}{\left(\dfrac{5.0 \times 10^{-3} \text{ kg}}{2.2 \text{ m}}\right)}} = \boxed{185\frac{\text{m}}{\text{s}}}$$

REFLECT

The propagation speed is how fast the wave travels down the rope, not how fast a small piece of the rope moves up and down.

Get Help: P'Cast 13.1 – Correct Units for Propagation Speed

13.61

SET UP

The bulk modulus and density of water are $B = 2.2 \times 10^9$ Pa and $\rho = 1000$ kg/m^3, respectively. We can use this information to find the speed of sound in water using $v_p = \sqrt{\dfrac{B}{\rho}}$.

SOLVE

$$v_p = \sqrt{\frac{B}{\rho}} = \sqrt{\frac{2.2 \times 10^9 \text{ Pa}}{\left(1000\dfrac{\text{kg}}{\text{m}^3}\right)}} = \boxed{1483\frac{\text{m}}{\text{s}}}$$

REFLECT

For comparison, the speed of sound in air is 343 m/s.

Get Help: P'Cast 13.2 – Propagation Speeds

13.67

SET UP

We are shown four different scenarios where square waveforms of different widths and amplitudes are approaching one another (see figure). When they overlap in time and space, they will interfere with one another. If the waveforms are both upright, they will constructively interfere; if the waveforms are inverted relative to one another, they will destructively interfere.

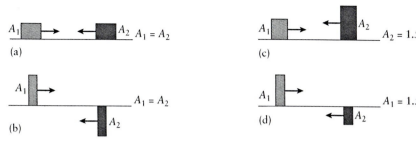

Figure 13-3 Problem 67

SOLVE

Part a) The resulting waveform will have the same width as the incoming waveforms but an amplitude of $A = A_1 + A_2 = \boxed{2A_1}$.

Part b) The resulting waveform will have an amplitude of $A = A_1 - A_2 = \boxed{0}$.

Part c) The resulting waveform will have the same width as the incoming waveforms but an amplitude of $A = A_1 + A_2 = \boxed{2.5A_1}$.

Part d) The resulting waveform will have the same width as the incoming waveforms but an amplitude of $A = A_1 - A_2 = \boxed{A_1/3}$.

REFLECT

The waveforms will pass through one another unaffected. Only in the exact moment when they are coincident will we observe another square waveform with the amplitudes we found.

13.71

SET UP

Two speakers are 6 m apart and playing the same tone ($f = 171.5$ Hz) in phase. The speed of sound ($v = 343$ m/s) divided by the frequency of the tone will give us its wavelength. In order to determine which of the labeled points (A−E) will experience fully constructive interference, we need to use geometry to determine the path length difference between the two waves. The path length for each wave is equal to the hypotenuse of a triangle made by the 6-m distance between the speakers and the line of points and the distance to the point. Keep in mind that each of the labeled points is 1 m from its neighbors. If the quotient of the path length difference and the wavelength of the tone is an integer, then that point will experience fully constructive interference; if not, there will be partial destructive interference.

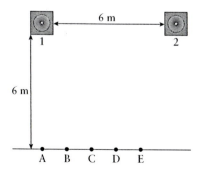

Figure 13-4 Problem 71

SOLVE

Wavelength:

$$\lambda = \frac{v}{f} = \frac{\left(343\frac{m}{s}\right)}{171.5 \text{ Hz}} = 2 \text{ m}$$

Point A)

$$\Delta_{pl} = D_2 - D_1 = (\sqrt{(6 \text{ m})^2 + (6 \text{ m})^2}) - (6 \text{ m}) = 2.49 \text{ m}$$

$$\Delta_{pl} = n\lambda$$

$$n = \frac{\Delta_{pl}}{\lambda} = \frac{2.49 \text{ m}}{2 \text{ m}} = 1.25$$

Since n is not an integer, point A $\boxed{\text{will not experience constructive interference}}$.

Point B)

$$\Delta_{pl} = D_2 - D_1 = (\sqrt{(5 \text{ m})^2 + (6 \text{ m})^2}) - (\sqrt{(1 \text{ m})^2 + (6 \text{ m})^2}) = 1.73 \text{ m}$$

$$\Delta_{pl} = n\lambda$$

$$n = \frac{\Delta_{pl}}{\lambda} = \frac{1.73 \text{ m}}{2 \text{ m}} = 0.865$$

Since n is not an integer, point B $\boxed{\text{will not experience constructive interference}}$.

Point C)

$$\Delta_{pl} = D_2 - D_1 = (\sqrt{(4 \text{ m})^2 + (6 \text{ m})^2}) - (\sqrt{(2 \text{ m})^2 + (6 \text{ m})^2}) = 0.887 \text{ m}$$

$$\Delta_{pl} = n\lambda$$

$$n = \frac{\Delta_{pl}}{\lambda} = \frac{0.887 \text{ m}}{2 \text{ m}} = 0.444$$

Since n is not an integer, point C $\boxed{\text{will not experience constructive interference}}$.

Point D)

$$\Delta_{pl} = D_2 - D_1 = (\sqrt{(3 \text{ m})^2 + (6 \text{ m})^2}) - (\sqrt{(3 \text{ m})^2 + (6 \text{ m})^2}) = 0 \text{ m}$$

$$\Delta_{pl} = n\lambda$$

$$n = \frac{\Delta_{pl}}{\lambda} = \frac{0 \text{ m}}{2 \text{ m}} = 0$$

Since n is an integer, point D $\boxed{\text{will experience constructive interference}}$.

Point E)

$$\Delta_{pl} = D_2 - D_1 = (\sqrt{(2 \text{ m})^2 + (6 \text{ m})^2}) - (\sqrt{(4 \text{ m})^2 + (6 \text{ m})^2}) = -0.887 \text{ m}$$

$$\Delta_{pl} = n\lambda$$

$$n = \frac{\Delta_{pl}}{\lambda} = \frac{-0.887 \text{ m}}{2 \text{ m}} = -0.444$$

Since n is not an integer, point E $\boxed{\text{will not experience constructive interference}}$.

REFLECT

The speakers are symmetric about a vertical line through point D, which means the distances from each speaker to that point will be equal.

Get Help: P'Cast 13.4 – Two Elephants

13.75

SET UP

A string with a length $L = 1.25$ m and mass $m = 0.0548$ kg has a tension of $T = 200$ N. It is fixed at both ends. We want to calculate the frequencies for the first four harmonics. The speed of the waves on the string is equal to $v = \sqrt{\dfrac{T}{\mu}}$, where μ is the linear mass density. The allowed frequencies are given by $f = \dfrac{v}{\lambda} = \dfrac{vn}{2L}$.

SOLVE

Part a)

Standing wave frequencies:

$$f = \frac{vn}{2L} = \frac{n}{2L}\sqrt{\frac{T}{\mu}} = \frac{n}{2L}\sqrt{\frac{T}{\left(\dfrac{m}{L}\right)}} = \frac{n}{2}\sqrt{\frac{T}{Lm}}$$

First harmonic ($n = 1$):

$$f_1 = \frac{1}{2}\sqrt{\frac{200 \text{ N}}{(1.25 \text{ m})(0.0548 \text{ kg})}} = \boxed{27.0 \text{ Hz}}$$

Second harmonic ($n = 2$):

$$f_1 = \sqrt{\frac{200 \text{ N}}{(1.25 \text{ m})(0.0548 \text{ kg})}} = \boxed{54.0 \text{ Hz}}$$

Third harmonic ($n = 3$):

$$f_1 = \frac{3}{2}\sqrt{\frac{200 \text{ N}}{(1.25 \text{ m})(0.0548 \text{ kg})}} = \boxed{81.0 \text{ Hz}}$$

Fourth harmonic ($n = 4$):

$$f_1 = 2\sqrt{\frac{200 \text{ N}}{(1.25 \text{ m})(0.0548 \text{ kg})}} = \boxed{108 \text{ Hz}}$$

Part b)

Figure 13-5 Problem 75

First harmonic = purple ($n = 1$)

Second harmonic = green ($n = 2$)

Third harmonic = red ($n = 3$)

Fourth harmonic = blue ($n = 4$)

REFLECT

The frequencies of the higher harmonics are integer multiples of the fundamental frequency, $f_n = n f_1$.

Get Help: Picture It – Harmonic Synthesis
P'Cast 13.5 – Fundamental

13.81

SET UP

A steel string ($\rho_{steel} = 7800 \text{ kg/m}^3$) has an unstretched length of $L = 0.325$ m and a diameter of $D = 0.25 \times 10^{-3}$ m. By setting the two expressions for the speed of a wave on a string, $v = \sqrt{\frac{T}{\mu}}$ and $v = \lambda f$, equal to one another, we can solve for the tension in the string. We can rewrite the linear mass density in terms of the density of steel, the volume of the string, and L. The fundamental frequency of this string is $f = 660$ Hz and the fundamental wavelength is $2L$.

SOLVE

$$v = \sqrt{\frac{T}{\mu}} = \sqrt{\frac{T}{\left(\frac{M}{L}\right)}} = \sqrt{\frac{TL}{\rho V}} = \sqrt{\frac{TL}{\rho\left(\pi\left(\frac{D}{2}\right)^2 L\right)}} = \sqrt{\frac{4T}{\rho \pi D^2}} = \lambda_1 f_1 = (2L)f$$

$$T = \pi L^2 f^2 \rho D^2 = \pi (0.325 \text{ m})^2 (660 \text{ Hz})^2 \left(7800\frac{\text{kg}}{\text{m}^3}\right)(0.25 \times 10^{-3} \text{ m})^2 = \boxed{70.5 \text{ N}}$$

REFLECT

The Young's modulus of steel is 200×10^9 Pa. The strain due to the tensile stress on the string is 0.7%, so we can safely assume that the stretched length of the string is approximately equal to its unstretched length.

Get Help: Interactive Exercise – Wave on a Rope

13.83

SET UP

Two successive harmonics of an organ pipe are 228.6 Hz and 274.3 Hz. We are asked to determine whether the pipe is open on both ends or open on one end and closed at the other end. Taking the ratio of these two frequencies will tell us the values of n to which these frequencies correspond. Only odd harmonics are allowed in an open–closed pipe, which means the integers in this ratio must differ by 2. An open–open pipe has harmonics with successive values of n.

SOLVE

Open–closed:

$$\frac{f_{n+2}}{f_n} = \frac{(n+2)f_1}{nf_1} = \frac{n+2}{n}$$

$$\frac{274.3 \text{ Hz}}{228.6 \text{ Hz}} = \frac{6}{5} \overset{?}{=} \frac{n+2}{n}$$

The integers in this fraction must both be odd if the pipe is open on one end and closed on the other end. Therefore, it is $\boxed{\text{not an open–closed pipe}}$.

Open–open:

$$\frac{f_{n+1}}{f_n} = \frac{(n+1)f_1}{nf_1} = \frac{n+1}{n}$$

$$\frac{274.3 \text{ Hz}}{228.6 \text{ Hz}} = \frac{6}{5} \overset{?}{=} \frac{n+1}{n}$$

The integers in this fraction must be successive if the pipe is open on both ends, which 5 and 6 are. Therefore, it is $\boxed{\text{an open–open pipe}}$.

REFLECT

We didn't have to calculate out both cases. Once we divided the frequencies and saw that the fraction was 6/5, we knew right away that it had to be an open–open pipe.

13.89

SET UP

The longest pipe in a pipe organ has a length of 9.75 m and the shortest is 0.0191 m. Assuming the speed of air is 343 m/s, we can find the range of the organ by calculating the fundamental frequencies of these pipes. We need to treat them as both open–open pipes $\left(f_1 = \dfrac{v}{2L} \right)$ and open–closed pipes $\left(f_1 = \dfrac{v}{4L} \right)$.

SOLVE

Fundamental frequencies for open–open pipes:

$$f_{\text{low}} = \frac{v}{2L} = \frac{\left(343 \dfrac{\text{m}}{\text{s}} \right)}{2(9.75 \text{ m})} = \boxed{17.6 \text{ Hz}}$$

$$f_{high} = \frac{v}{2L} = \frac{\left(343\frac{m}{s}\right)}{2(0.0191\ m)} = \boxed{8980\ Hz}$$

Fundamental frequencies for open–closed pipes:

$$f_{low} = \frac{v}{4L} = \frac{\left(343\frac{m}{s}\right)}{4(9.75\ m)} = \boxed{8.79\ Hz}$$

$$f_{high} = \frac{v}{4L} = \frac{\left(343\frac{m}{s}\right)}{4(0.0191\ m)} = \boxed{4490\ Hz}$$

REFLECT

The largest range of frequencies occurs when the long pipe acts as an open–closed pipe and the short pipe acts as an open–open one.

Get Help: P'Cast 13.6 – Tree-hole Frog

13.93

SET UP

A guitar string under 100 N of tension is supposed to have a frequency of 110 Hz. But, when played at the same time as a reference tone at 110 Hz, beats at 2 Hz are heard. The tension in the string is decreased, and the beat frequency increases, which means the string is becoming more out of tune. For a guitar, the length of the string and, thus, the wavelength are fixed. Therefore, the frequency is proportional to the square root of the tension. When the tension is decreased, the frequency of the sound will also decrease. Because the beat frequency increases when the tension in the string decreases, the initial frequency of the guitar string must be below the desired frequency. Once we know the initial frequency of the guitar, we can set up a ratio between the tensions and the frequencies and solve for the final desired tension when the string is in tune.

SOLVE

Initial frequency:

$$f_{beats} = |f_2 - f_1|$$

$$f_2 = f_1 \pm f_{beats} = (110\ Hz) \pm (2\ Hz) = 112\ Hz\ or\ 108\ Hz$$

Because the beat frequency increases when the tension in the string decreases, the initial frequency of the guitar string must be below the desired frequency, or $f_1 = 108$ Hz.

Tension:

$$\frac{v_1}{v_2} = \frac{\lambda f_1}{\lambda f_2} = \frac{\sqrt{\frac{T_1}{\mu}}}{\sqrt{\frac{T_2}{\mu}}}$$

$$\frac{f_1}{f_2} = \sqrt{\frac{T_1}{T_2}}$$

$$T_2 = T_1\left(\frac{f_2}{f_1}\right)^2 = (100 \text{ N})\left(\frac{110 \text{ Hz}}{108 \text{ Hz}}\right)^2 = \boxed{103.7 \text{ N}}$$

REFLECT

We are told that decreasing the tension increases the beat frequency. Therefore, the tension should be higher than 100 N when the guitar is in tune.

13.97

SET UP

Rush hour traffic lasts for 4 hr each day. A nearby resident plans to convert this sound energy into a more useful form of energy by collecting the sound with a 1-m² microphone that has an efficiency of 30%. The sound level of the traffic at the location of the microphone is $\beta = 100$ dB. The definition of the sound level relates the intensity of the sound at a given position to the reference intensity, $I_0 = 10^{-12}$ W/m². The total power incident the microphone is equal to the intensity at the microphone multiplied by the surface area of the microphone. The total sound energy due to the traffic is equal to the power multiplied by the time interval the traffic is active, or 4 hr. Because the microphone only absorbs 30% of the total energy that hits it, we need to multiply the total energy by 0.30 to find the total energy collected by the microphone.

SOLVE

Power:

$$\beta = 10\log\left(\frac{I}{I_0}\right)$$

$$I = I_0 10^{\frac{\beta}{10}} = \left(10^{-12}\frac{\text{W}}{\text{m}^2}\right)10^{\frac{100}{10}} = 10^{-2}\frac{\text{W}}{\text{m}^2}$$

$$I = \frac{P}{A}$$

$$P = IA = \left(10^{-2}\frac{\text{W}}{\text{m}^2}\right)(1 \text{ m}^2) = 10^{-2} \text{ W}$$

Energy collected:

$$E_{\text{total}} = P\Delta t = (10^{-2} \text{ W})\left(4 \text{ hr} \times \frac{3600 \text{ s}}{1 \text{ hr}}\right) = 144 \text{ J}$$

$$E_{\text{collected}} = (0.30)E_{\text{total}} = (0.30)(144 \text{ J}) = \boxed{43 \text{ J}}$$

She would do better by completely replacing the plan with something else. Barring that, she could move the collector closer to the highway (above the sound-blocking barrier would be best) or make it much larger (much of the cost would be in electronics, not the collector). Focusing wouldn't do much in the horizontal direction as the source is diffuse, but vertical focusing reflectors could help.

REFLECT

An energy amount of 43 J is not very much. If this amount of energy were in the form of heat, it would increase the temperature of 1 kg of water by only 0.01 degrees C.

13.101

SET UP

We can use the formula we derived in Problem 13.100 to relate an increase in sound level to an increase in intensity.

SOLVE

Part a)

$$\Delta \beta = 10 \log\left(\frac{I_2}{I_1}\right) = 1 \text{ dB}$$

$$I_2 = I_1 10^{\frac{1}{10}} = \boxed{1.26 I_1}$$

Part b)

$$\Delta \beta = 10 \log\left(\frac{I_2}{I_1}\right) = 20 \text{ dB}$$

$$I_2 = I_1 10^2 = \boxed{100 I_1}$$

REFLECT

We would expect the intensity to increase if there is an increase in the sound level.

13.107

SET UP

A bicyclist is moving toward a sheer wall at an unknown speed v while holding a tuning fork ringing at $f_0 = 484$ Hz. The bicyclist detects a beat frequency of 6 Hz between the sound from the tuning fork and the sound reflecting off the wall. Because the bicycle is moving toward the wall, the reflected frequency will be higher than f_0. Once we know the reflected frequency, we can use $f'' = \left(\frac{v_s + v}{v_s - v}\right) f_0$ to calculate the speed of the bicycle. We'll assume the speed of sound in air to be 343 m/s.

SOLVE

Observed reflected frequency:

$$f_{\text{beats}} = 6 \text{ Hz} = |f'' - (484 \text{ Hz})|$$

The bicycle is moving toward the wall, so we expect the observed reflected frequency to be higher than the original frequency. Therefore:

$$f'' = (484 \text{ Hz}) + (6 \text{ Hz}) = 490 \text{ Hz}$$

Speed of the bicycle:

$$f'' = \left(\frac{v_s + v}{v_s - v}\right) f_0$$

$$\frac{f''}{f_0}(v_s - v) = v_s + v$$

$$\left(\frac{f''}{f_0} - 1\right) v_s = \left(\frac{f''}{f_0} + 1\right) v$$

$$v = \frac{\left(\dfrac{f''}{f_0} - 1\right)}{\left(\dfrac{f''}{f_0} + 1\right)} v_s = \frac{\left(\dfrac{490\ \text{Hz}}{484\ \text{Hz}} - 1\right)}{\left(\dfrac{490\ \text{Hz}}{484\ \text{Hz}} + 1\right)}\left(343\frac{\text{m}}{\text{s}}\right) = \boxed{2.11\frac{\text{m}}{\text{s}}}$$

REFLECT

A speed of 2.11 m/s is a little under 5 mph, which is a reasonable speed for a person cycling while trying to hold a tuning fork.

13.109

SET UP

A bat emits a high-pitched sound at $f = 50{,}000$ Hz while traveling at 10 m/s toward an insect that is traveling away from it. The bat observes the reflected wave that echoes off the insect at a frequency of $f_{\text{obs}} = 50{,}050$ Hz. The general Doppler shift is given by $f_{\text{obs}} = \left(\dfrac{v_s \pm v_{\text{obs}}}{v_s \mp v_{\text{src}}}\right)f$, where the top signs correspond to motion "toward" and the bottom signs correspond to motion "away." The bat acts as a moving source traveling toward the observer at $v_{\text{src}} = 10$ m/s. The insect acts as a moving observer traveling away from the source at v_{obs}. This means we'll use the top sign ($-$) for the source term and the bottom sign for the observer term ($-$).

SOLVE

$$f_{\text{obs}} = \left(\frac{v_s - v_{\text{obs}}}{v_s - v_{\text{src}}}\right)f$$

$$v_{\text{obs}} = v_s - \frac{f_{\text{obs}}}{f}(v_s - v_{\text{src}}) = \left(343\frac{\text{m}}{\text{s}}\right) - \frac{50{,}050\ \text{Hz}}{50{,}000\ \text{Hz}}\left(\left(343\frac{\text{m}}{\text{s}}\right) - \left(10\frac{\text{m}}{\text{s}}\right)\right) = \boxed{9.67\frac{\text{m}}{\text{s}}}$$

REFLECT

If the mosquito doesn't change its path, the bat traveling at 10 m/s in the same direction will catch up to it and assuredly eat it.

13.115

SET UP

The speed of sound in helium and air at 0 degrees Celsius is $v_{\text{He}} = 972$ m/s and $v_{\text{air}} = 331$ m/s, respectively. We can use the speed of sound and density of helium ($\rho_{\text{He}} = 0.1786$ kg/m^3) to find the bulk modulus of helium. A person producing a frequency of 0.500 kHz in air will produce a different frequency if his respiratory tract is filled with helium. Since the effective length of the resonator remains constant, the wavelengths of the standing waves will also be constant, regardless of the gas in the resonator. Therefore, the ratio of the speeds of sound in helium and air will equal the ratio of the frequencies in helium and air. The addition of helium to the respiratory tract will affect the mix of frequencies excited in the resonator and, thus, the sound produced.

SOLVE

Part a)

$$v_{\text{He}} = \sqrt{\frac{B_{\text{He}}}{\rho_{\text{He}}}}$$

$$B_{\text{He}} = \rho_{\text{He}} v_{\text{He}}^2 = \left(0.1786\frac{\text{kg}}{\text{m}^3}\right)\left(972\frac{\text{m}}{\text{s}}\right)^2 = \boxed{1.69 \times 10^5\ \text{Pa}}$$

Part b)

$$\frac{v_{He}}{v_{air}} = \frac{\lambda f_{He}}{\lambda f_{air}}$$

$$f_{He} = \left(\frac{v_{He}}{v_{air}}\right)f_{air} = \frac{\left(972\frac{m}{s}\right)}{\left(331\frac{m}{s}\right)}(0.500 \text{ kHz}) = \boxed{1.47 \text{ kHz}}$$

Part c) The parts of the voice resonator that are in the air all have a frequency increase from the substitution of some helium into the air; the parts of the resonator that are in the body do not. This produces a different mix of frequencies and a different general sound.

REFLECT

The speed of sound in sulfur hexafluoride (SF_6) at room temperature is about $0.44 v_{air}$. Introducing it into the respiratory tract will have the opposite effect of helium on the person's voice by making it deeper.

Get Help: P'Cast 13.2 – Propagation Speeds

13.117

SET UP

The frequency of a musical note that is exactly an octave higher than another note is twice that of the lower note; correspondingly, the frequency of a note that is lower by an octave is half as big. A note that is two octaves above concert A (440 Hz) will, therefore, be four times higher. The ratio of the frequency of middle C to a note one octave below it will be equal to the square root of the ratio between the tensions in the string while playing those notes.

SOLVE

Part a)

One octave:

$$A' = 2A$$

Two octaves:

$$A'' = 2A' = 2(2A) = 4A = 4(440 \text{ Hz}) = \boxed{1760 \text{ Hz}}$$

Part b)

$$\frac{f_{middle\ C}}{f_{octave\ below}} = \frac{\left(\dfrac{v_{middle\ C}}{\lambda}\right)}{\left(\dfrac{v_{octave\ below}}{\lambda}\right)} = \frac{\left(\sqrt{\dfrac{T_{middle\ C}}{\mu}}\right)}{\left(\sqrt{\dfrac{T_{octave\ below}}{\mu}}\right)} = \sqrt{\frac{T_{middle\ C}}{T_{octave\ below}}}$$

$$T_{octave\ below} = (T_{middle\ C})\left(\frac{f_{octave\ below}}{f_{middle\ C}}\right)^2 = (T)\left(\frac{f_{octave\ below}}{2f_{octave\ below}}\right)^2 = \boxed{\frac{T}{4}}$$

REFLECT

We can safely assume the length of the vibrating string on the viola does not change, which means the wavelengths of the standing wave will not change. The frequency of the note corresponds to the pitch that we hear.

13.123

SET UP

A sound level of 120 dB was recorded from 10-Hz sound waves emitted from a volcanic eruption. The intensity of the sound at the location of the detector is related both to the sound level $\left(\beta = 10\log\left(\dfrac{I}{I_0}\right)\right)$ and to the maximum displacement of the air molecules s_{max} $(I = 2\pi^2 \rho v_p f^2 s_{max}^2)$, where $\rho = 1.2$ kg/m³. We can also relate the intensity to the amount of energy the wave would deliver to a 2.0 m × 3.0 m wall located at the site of the detector within 1.0 min, $I = \dfrac{P}{A} = \dfrac{\Delta E}{A(\Delta t)}$; recall that power is energy transferred per time.

SOLVE

Part a)

Intensity:

$$\beta = 10\log\left(\frac{I}{I_0}\right)$$

$$I = \left(10^{\frac{\beta}{10}}\right)I_0$$

Maximum displacement:

$$I = 2\pi^2 \rho v_p f^2 s_{max}^2 = \left(10^{\frac{\beta}{10}}\right)I_0$$

$$s_{max} = \sqrt{\frac{\left(10^{\frac{\beta}{10}}\right)I_0}{2\pi^2 \rho v_p f^2}} = \sqrt{\frac{\left(10^{\frac{120}{10}}\right)\left(10^{-12}\dfrac{W}{m^2}\right)}{2\pi^2\left(1.2\dfrac{kg}{m^3}\right)\left(343\dfrac{m}{s}\right)(10\ Hz)^2}} = \boxed{0.0011\ m = 1.1\ mm}$$

Part b)

$$I = \frac{P}{A} = \frac{\Delta E}{A(\Delta t)} = \left(10^{\frac{\beta}{10}}\right)I_0$$

$$\Delta E = \left(10^{\frac{\beta}{10}}\right)I_0 A(\Delta t) = \left(10^{\frac{120}{10}}\right)\left(10^{-12}\frac{W}{m^2}\right)((2.0\ m)(3.0\ m))(60\ s) = \boxed{360\ J}$$

REFLECT

Since we don't know the distance between the volcano and the sound detector, we have to assume the wall is located at the same distance away.

13.129

SET UP

Two 40-W speakers, labeled A and B, are playing the same 857.5-Hz tone in phase. The speakers are 8 m apart, and speaker A is 6 m in front of you. In order to determine the intensity where you are, we need to determine whether or not there is constructive interference by calculating the path difference between the sound waves from A and B and compare it to the wavelength. If there is fully constructive interference, the total intensity is the sum of the intensity of each speaker at that position, where the intensity of each speaker is $I = \dfrac{P}{4\pi r^2}$. The sound level is equal to $\beta = 10\log\left(\dfrac{I_{total}}{I_0}\right)$.

Figure 13-7 Problem 129

SOLVE

Wavelength:

$$\lambda = \frac{v}{f} = \frac{\left(343\frac{m}{s}\right)}{857.5 \text{ Hz}} = 0.4 \text{ m}$$

Interference:

$$\Delta_{pl} = D_B - D_A = \left(\sqrt{(8 \text{ m})^2 + (6 \text{ m})^2}\right) - (6 \text{ m}) = (10 \text{ m}) - (6 \text{ m}) = 4 \text{ m}$$

$$\frac{\Delta_{pl}}{\lambda} = \frac{4 \text{ m}}{0.4 \text{ m}} = 10$$

which means there will be fully constructive interference.

Total Intensity:

$$I_{total} = I_A + I_B = \frac{P}{4\pi r_A^2} + \frac{P}{4\pi r_B^2} = \frac{P}{4\pi}\left(\frac{1}{r_A^2} + \frac{1}{r_B^2}\right) = \frac{40 \text{ W}}{4\pi}\left(\frac{1}{(6 \text{ m})^2} + \frac{1}{(10 \text{ m})^2}\right) = \boxed{0.12\frac{W}{m^2}}$$

Sound level:

$$\beta = 10\log\left(\frac{I_{total}}{I_0}\right) = 10\log\left(\frac{\left(0.12\frac{W}{m^2}\right)}{\left(10^{-12}\frac{W}{m^2}\right)}\right) = \boxed{111 \text{ dB}}$$

REFLECT

A sound level of 111 dB is equivalent to a car horn at a distance of 1 m.

Chapter 14
Thermodynamics I

Conceptual Questions

14.3 The warmer the day, the faster the molecules will move, so the sooner the smell will spread.

14.9 Part a) They are the same temperature.

Part b) The steam can cause a more severe burn because of its latent heat of vaporization.

14.17 The two metals must be chosen to have different thermal expansion coefficients. Because the metals are attached directly, thermal expansion will bend the bimetallic strip, which in turn pushes or pulls on the switch.

Multiple-Choice Questions

14.27 B (reduced to one-quarter its original value).

$$v^2 \propto T$$

$$\frac{v_{rms,\,2}}{v_{rms,\,1}} = \frac{1}{2} = \sqrt{\frac{T_2}{T_1}}$$

$$T_2 = \frac{1}{4}T_1$$

Get Help: P'Cast 14.3 – Helium Balloon

14.31 D (the temperature of each object will be the same). The two objects will eventually reach thermal equilibrium.

14.35 C (halved). The rate of heat transfer is inversely proportional to the thickness of the wall L.

$$\frac{H_2}{H_1} = \frac{\left(\dfrac{1}{L_2}\right)}{\left(\dfrac{1}{L_1}\right)} = \frac{L_1}{L_2} = \frac{L_1}{(2L_1)} = \frac{1}{2}$$

Estimation Questions

14.41 After about 10 min, your muscles will shut down and your heart will stop after about an hour.

14.47 About 10 times more heat is lost to radiation compared to conduction.

Problems

14.57

SET UP

There exists a numerical value that is the same on both the Celsius and Fahrenheit scales. The heat energy transfers must be constant regardless of the temperature scale used, but the degree units on the Celsius and Fahrenheit scales are different. The relationship to the temperature is a fixed amount of energy transfer and the degree units on the Celsius and Fahrenheit scales are not the same. While a plot of average kinetic energy per molecule in a gas versus temperature yields a straight line using either scale, the lines have a different slope when plotted using Fahrenheit temperatures or Celsius temperatures. Therefore, these lines must cross.

SOLVE

Special value where $T_C = T_F$:

$$T_C = \frac{5}{9}(T_F - 32) = \frac{5}{9}(T_C - 32)$$

$$\frac{4}{9}T_C = (-32)\left(\frac{5}{9}\right)$$

$$\boxed{T_C = -40 = T_F}$$

REFLECT

The Celsius and Kelvin scales will never share a numerical value.

14.61

SET UP

An unknown gas has a volume $V = 4.13$ L and pressure $P = 10$ atm at a temperature $T = 293$ K. We can use the ideal gas law, with $R = 0.0821\dfrac{\text{L} \cdot \text{atm}}{\text{mol} \cdot \text{K}}$, to find the number of moles of gas. The gas sample has a mass of 55 g; dividing this by the number of moles yields the molar mass of the unknown gas. We can use the molar mass and a periodic table to determine the identity of the unknown gas.

SOLVE

Number of moles:

$$PV = nRT$$

$$n = \frac{PV}{RT} = \frac{(10 \text{ atm})(4.13 \text{ L})}{\left(0.0821\dfrac{\text{L} \cdot \text{atm}}{\text{mol} \cdot \text{K}}\right)(293 \text{ K})} = 1.72 \text{ mol}$$

Molar mass:

$$\frac{55 \text{ g}}{1.72 \text{ mol}} = 32.0\frac{\text{g}}{\text{mol}}$$

This is the molar mass of $\boxed{O_2}$.

REFLECT

Be sure all of your units are consistent when performing your calculation. The pressure and volume are given in non-SI units, so it makes sense to use $R = 0.0821 \dfrac{\text{L} \cdot \text{atm}}{\text{mol} \cdot \text{K}}$.

14.65

SET UP

Water vapor (H_2O, molecular mass = 18 amu) and oxygen (O_2, molecular mass = 32 amu) are both present in the atmosphere. Since the gases are allowed to travel in three dimensions, the average kinetic energy of a single molecule of the gas is $K_{avg} = \dfrac{1}{2}mv^2 = \dfrac{3}{2}kT$, and the square root of v^2 is v_{rms}. We assume the temperature is constant.

SOLVE

Root-mean-square speed:

$$K_{avg} = \frac{1}{2}mv^2 = \frac{3}{2}kT$$

$$v_{rms} = \sqrt{v^2} = \sqrt{\frac{3kT}{m}}$$

Ratio:

$$\frac{v_{rms,\, water}}{v_{rms,\, oxygen}} = \frac{\left(\sqrt{\dfrac{3kT}{m_{water}}}\right)}{\left(\sqrt{\dfrac{3kT}{m_{oxygen}}}\right)} = \sqrt{\frac{m_{oxygen}}{m_{water}}} = \sqrt{\frac{32 \text{ amu}}{18 \text{ amu}}} = \sqrt{\frac{16}{9}} = \frac{4}{3}$$

REFLECT

The lighter molecule (water) has a higher v_{rms} at a given temperature, which makes sense.

Get Help: P'Cast 14.3 – Helium Balloon

14.71

SET UP

The mean free path for O_2 molecules at a temperature of 300 K and a pressure of 1.00 atm is 7.10×10^{-8} m. We can rearrange the expression for the mean free path, $\lambda = \dfrac{kT}{4\sqrt{2}\pi r^2 P}$, in order to solve for the radius of an oxygen molecule.

SOLVE

$$\lambda = \frac{kT}{4\sqrt{2}\pi r^2 P}$$

$$r = \sqrt{\frac{kT}{4\sqrt{2}\pi \lambda P}} = \sqrt{\frac{\left(1.38 \times 10^{-23}\dfrac{\text{J}}{\text{K}}\right)(300 \text{ K})}{4\sqrt{2}\pi(7.10 \times 10^{-8} \text{ m})\left(1.00 \text{ atm} \times \dfrac{1.01 \times 10^5 \text{ Pa}}{1 \text{ atm}}\right)}} = \boxed{1.80 \times 10^{-10} \text{ m}}$$

REFLECT

The radius of an oxygen atom is 0.6×10^{-10} m and the length of an oxygen–oxygen double bond is 1.21×10^{-10} m. If we model O_2 as two solid spheres 0.6×10^{-10} m in radius separated by 1.21×10^{-10} m, we will get a diameter of 3.61×10^{-10} m, which agrees well with our answer.

14.75

SET UP

The temperature of a silver pin that is 5.00 cm long decreases by 152 K. The coefficient of linear expansion of silver is 19.5×10^{-6} K. The final length of the pin after cooling is equal to the initial length L_0 plus the change in the length due to the temperature change, $\Delta L = \alpha L_0 \Delta T$.

SOLVE

$$L_f = L_0 + \Delta L = L_0 + (\alpha L_0 \Delta T) = L_0(1 + \alpha \Delta T)$$

$$= (5.00 \text{ cm})(1 + (19.5 \times 10^{-6} \text{ K}^{-1})(-152 \text{ K})) = \boxed{4.985 \text{ cm}}$$

REFLECT

We expect the pin to get smaller as it cools down, so $L_f < L_0$.

Get Help: Picture It – Linear Expansion
P'Cast 14.4 – Buckling Bridge

14.79

SET UP

A cylinder of solid aluminum has an initial length $L_0 = 5$ m and an initial radius $R_0 = 2$ cm. The temperature of the aluminum is increased by 25 K. The length and radius of the cylinder both increase independently due to the temperature increase according to $\Delta L = \alpha L_0 \Delta T$, where the coefficient of linear expansion of aluminum is 22.2×10^{-6} K^{-1}. Since the mass of the cylinder remains constant but the volume increases, the density of the aluminum cylinder will decrease. We can set up a ratio of the densities in order to calculate the factor by which the density decreases and the volume increases.

SOLVE

Part a)

$$\Delta L = \alpha L_0 \Delta T = (22.2 \times 10^{-6} \text{ K}^{-1})(5 \text{ m})(25 \text{ K}) = \boxed{0.00278 \text{ m}}$$

Part b)

Final radius:

$$R_f = R_0 + \Delta R = R_0 + (\alpha R_0 \Delta T) = R_0(1 + \alpha \Delta T)$$

$$= (2 \text{ cm})(1 + (22.2 \times 10^{-6} \text{ K}^{-1})(25 \text{ K})) = 2.00111 \text{ cm}$$

Change in density:

$$\frac{\rho_f}{\rho_0} = \frac{\left(\dfrac{M}{V_f}\right)}{\left(\dfrac{M}{V_0}\right)} = \frac{V_0}{V_f} = \frac{\pi R_0^2 L_0}{\pi R_f^2 L_f} = \frac{R_0^2 L_0}{R_f^2 L_f} = \frac{(2 \text{ cm})^2(5 \text{ m})}{(2.00111 \text{ cm})^2(5.00278 \text{ m})} = 0.998$$

The density decreases by 0.2% upon heating.

Part c)

$$\frac{V_f}{V_0} = \left(\frac{\rho_f}{\rho_0}\right)^{-1} = (0.998)^{-1} = 1.002$$

The volume increases by 0.2% upon heating.

REFLECT

The mass of the cylinder must remain constant since we are not adding or removing anything from it; all we are doing is increasing its size. We could have also used the approximate relationship for the thermal expansion in three dimensions: $V_f = V_0 + \Delta V = V_0(1 + 3\alpha\Delta T)$.

Get Help: Picture It – Linear Expansion
P'Cast 14.4 – Buckling Bridge

14.83

SET UP

You want to heat up 0.250 kg of water from 20 degrees C to 95 degrees C. The minimum amount of heat required to accomplish that is given by $Q = mc(\Delta T)$, where c is the specific heat of water $c = 4186\frac{J}{kg \cdot K}$.

SOLVE

$$\Delta T = \frac{1}{mc}Q$$

$$Q = mc(\Delta T) = (0.250 \text{ kg})\left(4186\frac{J}{kg \cdot K}\right)(75 \text{ K}) = \boxed{7.8 \times 10^4 \text{ J} = 78 \text{ kJ}}$$

REFLECT

A temperature difference of 75 degrees C is the same as a temperature difference of 75 K. Accordingly, sometimes you'll see the specific heat of water as $c = 4186\frac{J}{kg \cdot °C}$.

Get Help: P'Cast 14.5 – Camping

14.87

SET UP

The temperature of a lake was increased by 5 K when 1.7×10^{14} J was transferred to it. We can calculate the mass of the lake from $\Delta T = \frac{1}{mc}Q$ and the specific heat of the lake $\left(4186\frac{J}{kg \cdot K}\right)$.

SOLVE

$$\Delta T = \frac{1}{mc}Q$$

$$m = \frac{Q}{c\Delta T} = \frac{1.7 \times 10^{14} \text{ J}}{\left(4186\frac{J}{kg \cdot K}\right)(5 \text{ K})} = \boxed{8.1 \times 10^9 \text{ kg}}$$

REFLECT

The lake has a volume of approximately 8.1×10^9 L. For a comparison, the volume of Lake Superior, the largest of the Great Lakes, is 1.2×10^{16} L.

14.93

SET UP

Each stroke of a hacksaw supplies 30 J of heat to a 20-g steel bolt. We can calculate the total heat necessary to raise the temperature of the bolt by 60 K from $Q = mc\Delta T$, where $c = 452 \dfrac{J}{kg \cdot K}$. The total heat divided by the heat supplied per stroke will give the total number of strokes required.

SOLVE

Total heat required:

$$Q = mc\Delta T = (0.020 \text{ kg})\left(452\frac{J}{kg \cdot K}\right)(60 \text{ K}) = 542.4 \text{ J}$$

Number of strokes:

$$n_{strokes} = \frac{\text{Total heat}}{\text{Heat supplied by one stroke}} = \frac{542.4 \text{ J}}{30 \text{ J}} = 18.1 = \boxed{19 \text{ strokes}}$$

REFLECT

The 19th stroke will cause an overall temperature increase of more than 60 K, but 18 strokes will only increase the temperature by 59.7 K.

Get Help: P'Cast 14.5 – Camping

14.97

SET UP

The total heat required to melt 0.400-kg of copper initially at 293 K is equal to the sum of the heat necessary to raise the temperature of the sample to the melting point ($T_f = 1357$ K) and the heat necessary to convert the entire solid into liquid. The heat necessary to increase the temperature of the object is given by $Q = mc\Delta T$, and the heat necessary to convert from solid to liquid is related to the latent heat of fusion, $Q = mL_F$. The specific heat of copper is $c = 387 \dfrac{J}{kg \cdot K}$, and the latent heat of fusion for copper in this problem is $L_F = 205{,}350 \dfrac{J}{kg}$.

SOLVE

Increase temperature to melting point:

$$Q = mc\Delta T = (0.400 \text{ kg})\left(387\frac{J}{kg \cdot K}\right)(1064 \text{ K}) = 164{,}707 \text{ J}$$

Melting the copper:

$$Q = mL_F = (0.400 \text{ kg})\left(205{,}350\frac{J}{kg}\right) = 82{,}140 \text{ J}$$

Total heat:

$$Q_{total} = (164{,}707 \text{ J}) + (82{,}140 \text{ J}) = 246{,}847 \text{ J} = \boxed{247 \text{ kJ}}$$

REFLECT

We have to increase the temperature of the copper by over 1300 K so it makes sense that most of the heat added goes toward this process rather than melting.

Get Help: P'Cast 14.6 − Melting Ice

14.101

SET UP

We want to see which sample of iron, copper, or water will melt the most from a huge block of ice. The iron sample has a mass of $m_{Fe} = 50$ g and an initial temperature of 120 degrees C; the copper sample has a mass of $m_{Cu} = 60$ g and an initial temperature of 150 degrees C; and the water has a mass of $m_{water} = 30$ g and an initial temperature of 40 degrees C. Assuming the sample and the ice form a thermally isolated system, the amount of heat leaving the sample is equal to the heat gained by the ice due to conservation of energy. The total heat lost by the sample is equal to the heat lost to lower its temperature from its initial value to 0 degrees C. The heat gained by the ice goes into heating it from −5 degrees C to 0 degrees C and then melting the ice. The heat necessary to change the temperature of an object is given by $Q = mc\Delta T$, and the heat necessary to convert from liquid to solid is related to the latent heat of fusion, $Q = mL_F$. The specific heats of iron, copper, liquid water, and ice are

$c_{Fe} = 452\dfrac{\text{J}}{\text{kg} \cdot \text{K}}$, $c_{Cu} = 387\dfrac{\text{J}}{\text{kg} \cdot \text{K}}$, $c_{water} = 4186\dfrac{\text{J}}{\text{kg} \cdot \text{K}}$, and $c_{ice} = 2093\dfrac{\text{J}}{\text{kg} \cdot \text{K}}$, respectively. The

latent heat of fusion for water is $L_F = 334 \times 10^3 \dfrac{\text{J}}{\text{kg}}$. We can calculate the mass of ice melted

in each case and compare these amounts to determine which sample melts the most.

SOLVE

Iron:

$$Q_{ice, \, -5°C \to 0°C} + Q_{ice \to water} + Q_{Fe, \, 120°C \to 0°C} = 0$$

$$m_{ice} c_{ice} \Delta T_{-5°C \to 0°C} + m_{ice} L_F + m_{Fe} c_{Fe} \Delta T_{120°C \to 0°C} = 0$$

$$m_{ice} = \frac{-m_{Fe} c_{Fe} \Delta T_{120°C \to 0°C}}{c_{ice} \Delta T_{-5°C \to 0°C} + L_F} = \frac{-(50 \text{ g})\left(452\dfrac{\text{J}}{\text{kg} \cdot \text{K}}\right)(-120 \text{ K})}{\left(2093\dfrac{\text{J}}{\text{kg} \cdot \text{K}}\right)(5 \text{ K}) + \left(334 \times 10^3 \dfrac{\text{J}}{\text{kg}}\right)} = \boxed{7.87 \text{ g}}$$

Copper:

$$Q_{ice, \, -5°C \to 0°C} + Q_{ice \to water} + Q_{Cu, \, 150°C \to 0°C} = 0$$

$$m_{ice} c_{ice} \Delta T_{-5°C \to 0°C} + m_{ice} L_F + m_{Cu} c_{Cu} \Delta T_{150°C \to 0°C} = 0$$

$$m_{ice} = \frac{-m_{Cu} c_{Cu} \Delta T_{150°C \to 0°C}}{c_{ice} \Delta T_{-5°C \to 0°C} + L_F} = \frac{-(60 \text{ g})\left(387\dfrac{\text{J}}{\text{kg} \cdot \text{K}}\right)(-150 \text{ K})}{\left(2093\dfrac{\text{J}}{\text{kg} \cdot \text{K}}\right)(5 \text{ K}) + \left(334 \times 10^3 \dfrac{\text{J}}{\text{kg}}\right)} = \boxed{10.1 \text{ g}}$$

Water:

$$Q_{\text{ice}, -5°C \to 0°C} + Q_{\text{ice} \to \text{water}} + Q_{\text{water}, 40°C \to 0°C} = 0$$

$$m_{\text{ice}} c_{\text{ice}} \Delta T_{-5°C \to 0°C} + m_{\text{ice}} L_F + m_{\text{water}} c_{\text{water}} \Delta T_{40°C \to 0°C} = 0$$

$$m_{\text{ice}} = \frac{-m_{\text{water}} c_{\text{water}} \Delta T_{40°C \to 0°C}}{c_{\text{ice}} \Delta T_{-5°C \to 0°C} + L_F} = \frac{-(30 \text{ g})\left(4186 \dfrac{\text{J}}{\text{kg} \cdot \text{K}}\right)(-40 \text{ K})}{\left(2093 \dfrac{\text{J}}{\text{kg} \cdot \text{K}}\right)(5 \text{ K}) + \left(334 \times 10^3 \dfrac{\text{J}}{\text{kg}}\right)} = \boxed{14.6 \text{ g}}$$

The $\boxed{\text{water will melt the most ice}}$.

REFLECT

Since we assume the ice block was really large, we do not have to worry about the sample and melted ice reaching an equilibrium temperature higher than 0 degrees C. Even though the mass of the water was the least, it melted the most ice mainly due to water's large specific heat.

14.105

SET UP

A star radiates 1000 times more power than our Sun, but its temperature is only 70% of the Sun's temperature. The radiated power is proportional to both the surface area and the temperature to the fourth power: $P = \sigma \varepsilon A T^4$. We can set up a ratio between the power radiated by each star in order to calculate the radius of the star in terms of the radius of the Sun, $R_{\text{Sun}} = 6.96 \times 10^8$ m. We assume that both stars are perfect emitters and that they are both perfectly spherical.

SOLVE

$$\frac{P_{\text{star}}}{P_{\text{Sun}}} = \frac{\sigma \varepsilon A_{\text{star}} T_{\text{star}}^4}{\sigma \varepsilon A_{\text{star}} T_{\text{Sun}}^4} = \frac{(4\pi R_{\text{star}}^2) T_{\text{star}}^4}{(4\pi R_{\text{Sun}}^2) T_{\text{Sun}}^4} = \frac{R_{\text{star}}^2 T_{\text{star}}^4}{R_{\text{Sun}}^2 T_{\text{Sun}}^4}$$

$$R_{\text{star}} = R_{\text{Sun}} \sqrt{\frac{P_{\text{star}} T_{\text{Sun}}^4}{P_{\text{Sun}} T_{\text{star}}^4}} = R_{\text{Sun}} \sqrt{\frac{(1000 P_{\text{Sun}}) T_{\text{Sun}}^4}{P_{\text{Sun}} (0.7 T_{\text{Sun}})^4}} = (6.96 \times 10^8 \text{ m}) \sqrt{\frac{1000}{(0.7)^4}} = \boxed{4.5 \times 10^{10} \text{ m}}$$

REFLECT

This is about 65 times larger than the radius of the Sun. Changing the temperature has a much larger effect on the power than changing the surface area since it is raised to the fourth power.

14.107

SET UP

A glass window is 0.30 m × 1.50 m and 1.2×10^{-3} m thick. We can calculate the rate of heat transfer through the window by conduction by $H = k\dfrac{A}{L}(T_H - T_C)$. The temperature difference across the window is 17 K, and the thermal conductivity of window glass is $k = 0.96 \dfrac{\text{W}}{\text{m} \cdot \text{K}}$.

SOLVE

$$H = k\frac{A}{L}(T_H - T_C) = \left(0.96\frac{W}{m \cdot K}\right)\frac{(0.30 \text{ m})(1.50 \text{ m})}{1.2 \times 10^{-3} \text{ m}}(17 \text{ K}) = \boxed{6100 \text{ W} = 6.1 \text{ kW}}$$

REFLECT

Unlike radiation, the rate of heat transfer by conduction depends on the temperature difference between two objects.

Get Help: Picture It – Thermal Current
P'Cast14.12 – Heat Loss through a Window

14.111

SET UP

The main constituents of the atmosphere on Mars are carbon dioxide (95.4%), nitrogen (2.7%), argon (1.6%), oxygen (0.13%), and carbon monoxide (0.007%). The atmospheric pressure on Mars is $P_{atm, Mars} = 0.675$ kPa. The partial pressure of a constituent gas is equal to the fraction of that gas in the mixture multiplied by the total atmospheric pressure.

SOLVE

Carbon dioxide:

$$P_{CO_2} = (0.954)P_{atm, Mars} = (0.954)(0.675 \text{ kPa}) = \boxed{0.644 \text{ kPa}}$$

Nitrogen:

$$P_{N_2} = (0.027)P_{atm, Mars} = (0.027)(0.675 \text{ kPa}) = \boxed{0.018 \text{ kPa}}$$

Argon:

$$P_{Ar} = (0.016)P_{atm, Mars} = (0.016)(0.675 \text{ kPa}) = \boxed{0.011 \text{ kPa}}$$

Oxygen:

$$P_{O_2} = (0.0013)P_{atm, Mars} = (0.0013)(0.675 \text{ kPa}) = \boxed{0.00088 \text{ kPa}}$$

Carbon monoxide:

$$P_{CO} = (0.0007)P_{atm, Mars} = (0.0007)(0.675 \text{ kPa}) = \boxed{0.0005 \text{ kPa}}$$

REFLECT

The sum of the partial pressures of *all* of the constituents of the atmosphere will equal $P_{atm, Mars}$. The above gases make up 99.9% of the atmosphere for a total pressure of 0.674 kPa.

14.117

SET UP

A brick wall is composed of 19.0-cm-long bricks ($\alpha_{brick} = 5.5 \times 10^{-6} \, °C^{-1}$) and 1.00-cm-long sections of mortar ($\alpha_{mortar} = 8.0 \times 10^{-6} \, °C^{-1}$). A 20-m-long wall will be made up of 100 of these repeating brick + mortar units. In order to calculate the expansion on this wall due to a temperature increase of 25 degrees C, we need to calculate the expansion of one brick and one piece of mortar due to this temperature increase. The total expansion of the wall will be equal to 100 times the expansion of one brick + mortar unit.

SOLVE

One brick + mortar unit:

$$\Delta L_{\text{brick}} = \alpha_{\text{brick}} L_{0,\,\text{brick}} \Delta T = (5.5 \times 10^{-6}\ {}^{\circ}\text{C}^{-1})(19.0\ \text{cm})(25{}^{\circ}\text{C}) = 0.00261\ \text{cm}$$

$$\Delta L_{\text{mortar}} = \alpha_{\text{mortar}} L_{0,\,\text{mortar}} \Delta T = (8.0 \times 10^{-6}\ {}^{\circ}\text{C}^{-1})(1.00\ \text{cm})(25{}^{\circ}\text{C}) = 0.000200\ \text{cm}$$

$$\Delta L_{\text{total}} = \Delta L_{\text{brick}} + \Delta L_{\text{mortar}} = (0.00261\ \text{cm}) + (0.000200\ \text{cm}) = 0.00281\ \text{cm}$$

There are 100 brick + mortar units in a 20-m-long wall, so:

$$\Delta L_{\text{wall}} = 100\Delta L_{\text{total}} = 100(0.00281\ \text{cm}) = 0.281\ \text{cm}$$

The length of the wall $\boxed{\text{increases by } 0.281\ \text{cm}}$.

REFLECT

A value of 2.81 mm seems to be a reasonable value for a wall to expand.

Get Help: Picture It – Linear Expansion
P'Cast14.4 – Buckling Bridge

14.121

SET UP

A person collects sound waves from traffic and converts them into an electrical signal used to heat 5 kg of water. After 7 days the temperature of the water increased 0.01 K. Using the specific heat of water $\left(c = 4186\dfrac{\text{J}}{\text{kg} \cdot \text{K}} \right)$, we can calculate the energy (in the form of heat) required to cause this temperature increase. The total acoustic power "caught" by the transducer is equal to this energy divided by 7 days.

SOLVE

Heat required to raise the temperature:

$$Q = mc\Delta T = (5\ \text{kg})\left(4186\dfrac{\text{J}}{\text{kg} \cdot \text{K}} \right)(0.01\ \text{K}) = 209.3\ \text{J}$$

Acoustic power collected:

$$P = \frac{E}{t} = \frac{209.3\ \text{J}}{\left(7\ \text{days} \times \dfrac{24\ \text{hr}}{1\ \text{day}} \times \dfrac{3600\ \text{s}}{1\ \text{hr}} \right)} = \boxed{3.46 \times 10^{-4}\ \text{W} = 346\ \mu\text{W}}$$

REFLECT

An efficiency of 100% means all of the acoustic energy is converted into electrical energy. In actuality, the efficiency would be (much) lower than 100%, so 346 μW is the minimum power that the transducer collects.

Get Help: P'Cast 14.5 – Camping

14.125

SET UP

The total heat required to condense 5×10^{-3} kg of helium initially at 30 degrees C is equal to the sum of the heat necessary to lower the temperature of the sample to the boiling point ($T_{\text{V}} = -268.93$ degrees C) and the heat necessary to convert the entire gas into liquid. The heat transfer necessary to decrease the temperature of the helium is given by $Q = mc\Delta T$, and the heat necessary to convert from gas to liquid is related to the latent heat of vaporization, $Q = mL_V$. The specific heat of helium is $c = 5193\dfrac{J}{kg \cdot °C}$, and the latent heat of vaporization for helium is $L_V = 21{,}000\dfrac{J}{kg}$.

SOLVE

$$Q_{total} = Q_{30°C \to -268.93°C} + Q_{gas \to liquid} = mc\Delta T_{30°C \to -268.93°C} + mL_V = m(c\Delta T_{30°C \to -268.93°C} + L_V)$$

$$= (5 \times 10^{-3} \text{ kg})\left(\left(5193\frac{J}{kg \cdot °C}\right)(-298.93°C) + \left(-21{,}000\frac{J}{kg}\right)\right) = \boxed{-7867 \text{ J}}$$

REFLECT

The majority of the heat lost by the helium goes into cooling the gas by almost 300 degrees C.

14.131

SET UP

A man has a surface area of $A = 2.1$ m^2 and a skin temperature of 303 K. Normally 80% of the food calories (that is, kcal) he eats are converted into heat. The man is in a 293-K room and wants to know how many food calories x he needs to eat in a 24-hr period in order to maintain his current body temperature. The net radiated energy is equal to the energy he radiates minus the energy the air radiates back into him. The power radiated is given by $P = \sigma \varepsilon A T^4$, where we assume $\varepsilon = 1$. The conversion between kcal and joules is 1 kcal = 4186 J.

SOLVE

$$E_{food} = E_{radiated} = P_{radiated}t$$

$$(0.80x)\left(\frac{4186 \text{ J}}{1 \text{ kcal}}\right) = (\sigma \varepsilon A T_{body}^4 - \sigma \varepsilon A T_{air}^4)t = (T_{body}^4 - T_{air}^4)\sigma \varepsilon A t$$

$$x = \frac{(T_{body}^4 - T_{air}^4)\sigma \varepsilon A t}{(0.80)\left(\dfrac{4186 \text{ J}}{1 \text{ kcal}}\right)}$$

$$= \frac{((303 \text{ K})^4 - (293 \text{ K})^4)\left(5.6703 \times 10^{-8}\dfrac{W}{m^2 \cdot K^4}\right)(1)(2.1 \text{ m}^2)\left(24 \text{ hr} \times \dfrac{3600 \text{ s}}{1 \text{ hr}}\right)}{(0.80)\left(\dfrac{4186 \text{ J}}{1 \text{ kcal}}\right)}$$

$$= \boxed{3250 \text{ kcal}}$$

A value of 3250 kcal seems $\boxed{\text{reasonable}}$ for a man who is 1.88m (6 ft. 2 in) tall with a mass of 80 kg (176 lb).

REFLECT

The recommended caloric intake for an adult male is around 2550 kcal. It would make sense that additional, more complicated processes are occurring in the body than our simple model takes into account.

14.137

SET UP

A method for detecting planets (radius r) orbiting a larger star (radius R) is to monitor the brightness of the star. If a planet passes in front of a star, it will block some of the emitted light and the star will appear less bright. We assume the planet does not radiate visible light. To determine the factor by which the intensity of the light is reduced when a planet is in front of the star, we can take a ratio between the change in intensity $\Delta I = I_0 - I_p$, where I_0 is the initial intensity and I_p is the intensity when the planet is in front of the star, and the initial intensity. The intensity for any observer is given by the power emitted divided by the area, which is related to the distance between the emitter and the observer; although these areas are the same, the effective emitted powers are not. The power emitted by the star is $P_0 = \sigma \varepsilon A_0 T^4$, where $A_0 = \pi R^2$ since the amount of light reaching an observer is equivalent to the power radiated by a disk of radius R. (Thinking about this another way, the observer can't see the light emitted from the backside of the star and effectively only sees light emitted from the cross section of the front, which is a circle.) The effective power emitted when the planet is in the way is $P_p = \sigma \varepsilon A_p T^4$, where $A_p = \pi R^2 - \pi r^2$ since the planet blocks the middle of the star.

By simplifying the ratio, we find that $\dfrac{\Delta I}{I_0} = \left(\dfrac{r}{R}\right)^2$. Using this expression and the radii of Jupiter ($r_{\text{Jupiter}} = 6.91 \times 10^7$ m), the Earth ($r_{\text{Earth}} = 6.38 \times 10^6$ m), and the Sun ($R_{\text{Sun}} = 6.96 \times 10^8$ m), we can look at the intensity reductions caused by the two planets.

SOLVE

Part a)

$$\frac{\Delta I}{I_0} = \frac{I_0 - I_p}{I_0} = 1 - \frac{I_p}{I_0} = 1 - \frac{\left(\dfrac{P_p}{A}\right)}{\left(\dfrac{P_p}{A}\right)} = 1 - \frac{P_p}{P_0} = 1 - \frac{\sigma \varepsilon A_p T^4}{\sigma \varepsilon A_0 T^4} = 1 - \frac{\pi R^2 - \pi r^2}{\pi R^2}$$

$$= 1 - \frac{R^2 - r^2}{R^2} = 1 - 1 + \frac{r^2}{R^2} = \boxed{\left(\frac{r}{R}\right)^2}$$

Part b)

Jupiter:

$$\left(\frac{\Delta I}{I_0}\right)_{\text{Jupiter}} = \left(\frac{r_{\text{Jupiter}}}{R_{\text{Sun}}}\right)^2 = \left(\frac{6.91 \times 10^7 \text{ m}}{6.96 \times 10^8 \text{ m}}\right)^2 = 0.0099 \approx \boxed{1\%}$$

Earth:

$$\left(\frac{\Delta I}{I_0}\right)_{\text{Earth}} = \left(\frac{r_{\text{Earth}}}{R_{\text{Sun}}}\right)^2 = \left(\frac{6.38 \times 10^6 \text{ m}}{6.96 \times 10^8 \text{ m}}\right)^2 = 8.4 \times 10^{-5} \approx \boxed{0.0084\%}$$

Part c) Jupiter should be easy to find by measuring the reduction in intensity. Earth is much harder to pick out but still possible if they are determined to keep looking.

REFLECT

Be careful not to confuse the different areas. For the intensity of a point source that creates spherical waves, we're interested in the surface area of the sphere with a radius equal to the distance between the source and the observer. The area we need for the radiated power is equal to the surface area (or effective area, in this case) of the emitter itself.

Chapter 15
Thermodynamics II

Conceptual Questions

15.1 Part a) Isothermal processes are those in which the temperature remains constant throughout. An example of an isothermal process is when an acid is slowly poured into a base and allowed to equilibrate with a water bath. The work done by isothermal processes (for an ideal gas) is given by $nRT\ln\left(\dfrac{V_f}{V_i}\right)$.

Part b) Adiabatic processes are those in which there is no heat lost or gained. An example of an adiabatic process is a bicycle pump that is used to increase the pressure and temperature of the air that is inside as it goes into the tire. The work done by an adiabatic process is simply the change in internal energy (because $Q = 0$).

Part c) Isobaric processes are those in which the pressure remains constant. An example of an isobaric process is a gas in a cylinder that is sealed with a movable piston. As the gas is heated, the volume will expand to balance out the increase in temperature but the pressure remains constant.

Part d) Isochoric processes are those in which the volume does not change. An example of an isochoric process is a gas inside a cylinder that cannot expand or contract. If you change the temperature of the gas, it will respond by changing the pressure. No work is done by isochoric processes.

15.5 Kinetic friction saps bulk kinetic energy and converts it into heat. Unless the engine was attempting to generate heat, the heat from friction is energy that won't be going to whatever it was the engine was supposed to accomplish.

15.9 It increases by the amount of the heat divided by the temperature for the cool object minus the amount of heat divided by the temperature for the hot object: $\Delta S = \dfrac{Q}{T_c} - \dfrac{Q}{T_h}$.

When there was a temperature difference, that difference could be harnessed to do work. Once the two objects have equilibrated, that method of extracting work is no longer possible because the difference no longer exists.

15.13 The irreversible processes are a bit like one-way streets—you can go around the block but not take a U-turn. More concretely, general heat loss is undone not by letting the heat back into the hot object but by finding something even hotter or by doing work on the object to heat it; the halting of a ball that rolled to a stop is not undone by letting the grass kick it back into motion but by going and fetching the ball. Note that the reversible adiabatic cooling does not let heat escape to the environment but rather stores the thermal energy as mechanical energy in another part of the system.

15.17 As the piston recedes, it will sap energy from the gas molecules striking it. This lowers the gas temperature, requiring it to be reheated by the hot reservoir—but that is then not isothermal. The slower the piston recedes, the less energy it saps, but if the engine is to actually run at all it does have to move. By not acting isothermally, there is heat, which means Q/T will be nonzero, and thus there is an entropy increase.

Multiple-Choice Questions

15.21 **B** (decreases). $Q = 0$ because it is thermally isolated. Work is being done by the gas on the piston, which means the internal energy and, thus, the temperature of the gas decreases.

Get Help: P'Cast 15.2 – Isothermal Expansion

15.25 **A** (pressure). An isothermal process is one in which the temperature remains constant throughout.

Get Help: Interactive Exercise – Heating Nitrogen
P'Cast 15.4 – Isobaric Expansion

15.29 **A** (two adiabatic processes and two isothermal processes). The processes are isothermal expansion, adiabatic expansion, isothermal compression, and adiabatic compression.

Get Help: Picture It – Carnot Engine
Interactive Exercise – Leaky System

Estimation Questions

15.33 If we treat the gas as ideal, then $U = \frac{3}{2}nRT$. One liter of oxygen is about 0.05 mol, since 22.4 L = 1 mol of gas. At $T = 20$ degrees C (or 293 K),

$$U_0 = \frac{3}{2}(0.05 \text{ mol})\left(8.31\frac{J}{\text{mol} \cdot K}\right)(293 \text{ K}) \approx 183 \text{ J}$$ and at $T = 100$ degrees C (or

373 K), $U_f = \frac{3}{2}(0.05 \text{ mol})\left(8.31\frac{J}{\text{mol} \cdot K}\right)(373 \text{ K}) \approx 232 \text{ J}$. Therefore, the change in internal energy is 40 K.

15.37 Assuming a 20-gallon tank and that 1 gallon of gasoline releases 125,000 BTU, the gasoline can release 2.64×10^9 J of total energy. Most internal combustion engines are around 30% efficient, so 7.9×10^8 J would be available to move the car. The force of the tires on the road is equal to the force of static friction, $F = \mu_s mg$. The coefficient of static friction is around 0.4 on wet pavement and 0.7 on dry pavement, so we'll choose $\mu_s = 0.5$ as a nice, round number. The work done by this force on a 1000-kg car is

$W = Fd = \mu_s mgd$. Solving for the distance, $d = \dfrac{7.9 \times 10^8 \text{ J}}{(0.5)(1000 \text{ kg})\left(9.8\dfrac{m}{s^2}\right)} = 160$ km, or 200 km to one significant figure.

15.41 Number of microstates:

$$W_n = \frac{(n_H + n_T)!}{n_H!n_T!}$$

Probability of any given macrostate:

$$P_n = \frac{W_n}{\sum\limits_{n} W_n}$$

Part a)

Macrostate	Microstates	Number of Microstates	Probability	Entropy
(5H,0T) n = 5	(HHHHH)	1	1/32	$k_B\ln(1)=0$
(4H,1T) n = 4	(HHHHT) (HHHTH) (HHTHH) (HTHHH) (THHHH)	5	5/32	$k_B\ln(5) = 1.61\,k_B$
(3H,2T) n = 3	(HHHTT) (HHTHT) (HTHHT) (THHHT) (HHTTH) (HTHTH) (THHTH) (HTTHH) (THTHH) (TTHHH)	10	10/32	$k_B\ln(10) = 2.30\,k_B$
(2H,3T) n = 2	(TTTHH) (TTHTH) (THTTH) (HTTTH) (TTHHT) (THTHT) (HTTHT) (THHTT) (HTHTT) (HHTTT)	10	10/32	$k_B\ln(10)=2.30\,k_B$
(1H,4T) n = 1	(TTTTH) (TTTHT) (TTHTT) (THTTT) (HTTTT)	5	5/32	$k_B\ln(5) = 1.61\,k_B$
(0H,5T) n = 0	(TTTTT)	1	1/32	$k_B\ln(1) = 0$

Part b)

Macrostate	#Microstates	Probability	Entropy
(20H,0T)	1	1/616,666 = 0.0000016%	$k_B\ln(1) = 0$
(19H,1T)	20	20/616,666 = 0.0000324%	$k_B\ln(20) = 3\,k_B$
(18H,2T)	190	190/616,666 = 0.000308%	$k_B\ln(190) = 5.25\,k_B$
(17H,3T)	1140	1140/616,666 = 0.00185%	$k_B\ln(1140) = 7.04\,k_B$
(16H,4T)	4845	4845/616,666 = 0.00786%	$k_B\ln(4845) = 8.49\,k_B$
(15H,5T)	15,504	15,504/616,666 = 0.0251%	$k_B\ln(15,504) = 9.65\,k_B$
(14H,6T)	38,760	38,760/616,666 = 0.0629%	$k_B\ln(38,760) = 10.6\,k_B$
(13H,7T)	77,520	77,520/616,666 = 0.126%	$k_B\ln(77,520) = 11.3\,k_B$
(12H,8T)	125,970	125,970/616,666 = 0.204%	$k_B\ln(125,970) = 11.7\,k_B$
(11H,9T)	167,960	167,960/616,666 = 0.272%	$k_B\ln(167,960) = 12.0\,k_B$
(10H,10T)	184,756	184,756/616,666 = 0.3%	$k_B\ln(184,756) = 12.1\,k_B$
(9H,11T)	167,960	167,960/616,666 = 0.272%	$k_B\ln(167,960) = 12\,k_B$
(8H,12T)	125,970	125,970/616,666 = 0.204%	$k_B\ln(125,970) = 11.7\,k_B$
(7H, 13T)	77,520	77,520/616,666 = 0.126%	$k_B\ln(77,520) = 11.3\,k_B$
(6H,14T)	38,760	38,760/616,666 = 0.0629%	$k_B\ln(38,760) = 10.6\,k_B$
(5H,15T)	15,504	15,504/616,666 = 0.0251%	$k_B\ln(15,504) = 9.65\,k_B$
(4H,16T)	4845	4845/616,666 = 0.00786%	$k_B\ln(4845) = 8.49\,k_B$
(3H,17T)	1140	1140/616,666 = 0.00185%	$k_B\ln(1140) = 7.04\,k_B$
(2H,18T)	190	190/616,666 = 0.000308%	$k_B\ln(190) = 5.25\,k_B$
(1H,19T)	20	20/616,666 = 0.0000324%	$k_B\ln(20) = 3\,k_B$
(0H,20T)	1	1/616,666 = 0.0000016%	$k_B\ln(1) = 0$

Part c) The distribution is peaked very sharply compared to its width, though in absolute terms (for the number of heads) far more broadly than the previous two distributions, just from the sheer numbers we're working with.

Problems

15.45

SET UP

A gas is heated such that it follows a vertical line path on a PV diagram. The initial pressure and volume are 1.0×10^5 Pa and 3.0 m^3, respectively. The final pressure and volume are 2.0×10^5 Pa and 3.0 m^3, respectively. The work done by the gas on its surroundings is equal to zero because the gas neither expands nor contracts (that is, its volume remains constant).

SOLVE

$$W = 0 \text{ because } \Delta V = 0$$

REFLECT

Another way of thinking about this is that the area under a horizontal line is zero. Since the volume remains constant throughout the process, this is an isochoric process.

Get Help: Picture It – P-V Diagram

15.49

SET UP

An expanding ideal gas does 8.8 kJ of work. The process occurs isothermally, which means the change in the internal energy of the gas is zero. From the first law of thermodynamics, this means the heat absorbed by the gas is equal to the work done by the gas.

SOLVE

$$\Delta U = 0 = Q - W$$

$$Q = W = \boxed{+8.8 \text{ kJ}}$$

REFLECT

The work done by a gas on its surroundings is considered to be positive, as is the heat absorbed by a gas.

15.53

SET UP

An ideal gas is contained in a vessel with fixed walls. While its pressure is decreased, 800 kJ of heat leaves the gas. Since the volume is fixed, the work done by the gas is equal to zero. The change in the internal energy of the gas is equal to Q in this case from the first law of thermodynamics. Heat leaving the system corresponds to a negative value, so $Q = -800$ kJ.

SOLVE

$$\Delta U = Q - W = (-800 \text{ kJ}) - 0 = \boxed{-800 \text{ kJ}}$$

REFLECT

This is an isochoric process because the volume is constant. Since the pressure of the gas decreases while the volume remains constant, the temperature of the gas must decrease. Therefore, we would expect heat to leave the system.

15.57

SET UP

An engine takes in 10 kJ and exhausts 6 kJ. The efficiency of the engine is equal to the difference in the energy it takes in and the energy it exhausts divided by the amount of energy it takes in: $e = \dfrac{W}{Q_H} = \dfrac{Q_H - |Q_C|}{Q_H}$.

SOLVE

$$e = \frac{W}{Q_H} = \frac{Q_H - |Q_C|}{Q_H} = \frac{(10 \text{ kJ}) - (6 \text{ kJ})}{10 \text{ kJ}} = 0.4 = \boxed{40\%}$$

REFLECT

Efficiency is most commonly given as a percentage.

15.61

SET UP

A furnace supplies 28 kW of thermal power at $T_H = 573$ K and exhausts energy at $T_C = 293$ K. The theoretical maximum efficiency of a heat engine is given by $e = 1 - \dfrac{T_C}{T_H}$. The actual efficiency is equal to the work output divided by the heat input, $e = \dfrac{W}{Q_H}$. We can divide the numerator and denominator by time, to convert the work and heat to work per second and thermal power, respectively. Assuming the actual efficiency is equal to the theoretical maximum efficiency, we can solve for the work per second $\dfrac{W}{\Delta t}$ given $\dfrac{Q_H}{\Delta t} = 28$ kW.

SOLVE

Theoretical maximum efficiency:

$$e = 1 - \frac{T_C}{T_H} = 1 - \frac{293 \text{ K}}{573 \text{ K}} = 0.4887$$

Work per second:

$$e = \frac{\left(\dfrac{W}{\Delta t}\right)}{\left(\dfrac{Q_H}{\Delta t}\right)}$$

$$\frac{W}{\Delta t} = e\left(\frac{Q_H}{\Delta t}\right) = (0.4887)(28 \text{ kW}) = \boxed{13.7 \text{ kW}}$$

REFLECT

This is the largest amount of work that can be expected since we've assumed the actual efficiency is equal to the theoretical maximum efficiency.

Get Help: Picture It – Carnot Engine

Interactive Exercise – Leaky System

15.65

SET UP

A refrigerator requires $W = 35$ J to remove $Q_C = 190$ J from its interior. The temperature of the surroundings is $T_H = 295$ K. The coefficient of performance of the refrigerator is given by $CP = \dfrac{Q_C}{W}$. We can then use the other expressions for the coefficient of performance, $CP = \dfrac{Q_C}{Q_H - Q_C}$ and $CP = \dfrac{T_C}{T_H - T_C}$, to calculate heat ejected to the surroundings Q_H and the internal temperature of the refrigerator, respectively.

SOLVE

Part a)

$$CP = \frac{Q_C}{W} = \frac{190 \text{ J}}{35 \text{ J}} = \boxed{5.4}$$

Part b)

$$CP = \frac{Q_C}{Q_H - Q_C}$$

$$(CP)(Q_H - Q_C) = Q_C$$

$$Q_H = \frac{Q_C + (CP)Q_c}{CP} = Q_C\left(\frac{1}{CP} + 1\right) = (190 \text{ J})\left(\frac{1}{5.4} + 1\right) = \boxed{225 \text{ J}}$$

Part c)

$$CP = \frac{T_C}{T_H - T_C}$$

$$(CP)(T_H - T_C) = T_C$$

$$T_C = \frac{(CP)T_H}{1 + CP} = \frac{(5.4)(295 \text{ K})}{1 + 5.4} = 249 \text{ K} = \boxed{-24^\circ\text{C}}$$

REFLECT

The fact the refrigerator cycle is reversible means it follows the Carnot cycle, which allows us to use the expression for the coefficient of performance based on the temperatures.

15.69

SET UP

The temperature of 4 g of helium gas is increased by 1 K at constant volume. We want to know the mass of oxygen gas that, using the same amount of heat, will experience a temperature increase of 1 K as well. The heat required to change the temperature of a gas at constant volume is given by $Q_V = nC_V\Delta T$. Using this expression, we can set up a ratio between the heats required by each gas, which are equal in this case, in terms of the numbers of moles and molar heat capacities at constant volume. The number of moles is equal to the mass divided by the molar mass of the gas. The data we'll need are the molar masses and the molar heat capacities at constant volume for both helium and oxygen: $MM_{He} = 4\dfrac{g}{mol}$, $MM_{O_2} = 32\dfrac{g}{mol}$, $C_{V, He} = 12.5\dfrac{J}{mol \cdot K}$, and $C_{V, O_2} = 21.1\dfrac{J}{mol \cdot K}$.

SOLVE

$$\frac{Q_{V, O_2}}{Q_{V, He}} = \frac{n_{O_2}C_{V, O_2}\Delta T}{n_{He}C_{V, He}\Delta T}$$

$$1 = \frac{\left(\dfrac{m_{O_2}}{MM_{O_2}}\right)C_{V, O_2}}{\left(\dfrac{m_{He}}{MM_{He}}\right)C_{V, He}} = \left(\frac{m_{O_2}}{m_{He}}\right)\left(\frac{MM_{He}}{MM_{O_2}}\right)\left(\frac{C_{V, O_2}}{C_{V, He}}\right)$$

$$m_{O_2} = m_{He}\left(\frac{MM_{O_2}}{MM_{He}}\right)\left(\frac{C_{V, He}}{C_{V, O_2}}\right) = (4\text{ g})\frac{\left(32\dfrac{g}{mol}\right)\left(12.5\dfrac{J}{mol \cdot K}\right)}{\left(4\dfrac{g}{mol}\right)\left(21.1\dfrac{J}{mol \cdot K}\right)} = \boxed{19\text{ g}}$$

REFLECT

This is about 0.6 mol of oxygen. For constant heat and temperature change, the product of the number of moles and the molar heat capacity is constant. Since the molar heat capacity of oxygen is larger than that of helium, the number of moles of oxygen must be less than the number of moles of helium under these circumstances.

15.73

SET UP

The temperature of air increases from $T_i = 300$ K to $T_f = 1130$ K. An ideal gas that undergoes an adiabatic expansion or compression obeys $TV^{\gamma-1} = $ constant. We can use this expression and $\gamma_{air} = 1.4$ to solve for the ratio of initial volume to final volume.

SOLVE

$$T_i V_i^{\gamma-1} = T_f V_f^{\gamma-1}$$

$$\frac{T_f}{T_i} = \frac{V_i^{\gamma-1}}{V_f^{\gamma-1}} = \left(\frac{V_i}{V_f}\right)^{\gamma-1}$$

$$\frac{V_i}{V_f} = \left(\frac{T_f}{T_i}\right)^{\frac{1}{\gamma-1}} = \left(\frac{1130 \text{ K}}{300 \text{ K}}\right)^{\frac{1}{1.4-1}} = \boxed{27.5}$$

REFLECT

The final temperature is larger, so the adiabatic process will be compression. Therefore, $V_f < V_i$.

15.77

SET UP

A sample of ice absorbs 6.68×10^4 J of heat at 273 K as it melts. The minimum entropy change of the system is given by $\Delta S = \frac{Q}{T}$.

SOLVE

$$\Delta S = \frac{Q}{T} = \frac{6.68 \times 10^4 \text{ J}}{273 \text{ K}} = \boxed{245 \frac{\text{J}}{\text{K}}}$$

REFLECT

The entropy change is equal to $\frac{Q}{T}$ only in the case of a reversible process. We do not need to use the mass of the ice since we are given the amount of heat absorbed.

15.81

SET UP

A car ($m = 1800$ kg) is traveling at a speed of $v_i = 80$ km/hr when it crashes into a wall. The temperature of the air is 300 K. The entropy change of the universe due to an irreversible process is equal to the total energy transferred divided by the temperature, $\Delta S = \frac{Q}{T}$.

SOLVE

$$\Delta S = \frac{Q}{T} = \frac{\left(\frac{1}{2}mv_i^2\right)}{T} = \frac{\left(\frac{1}{2}\right)(1800 \text{ kg})\left(80\frac{\text{km}}{\text{hr}} \times \frac{1000 \text{ m}}{1 \text{ km}} \times \frac{1 \text{ hr}}{3600 \text{ s}}\right)^2}{300 \text{ K}} = \boxed{1500\frac{\text{J}}{\text{K}} = 1.5\frac{\text{kJ}}{\text{K}}}$$

REFLECT

We're assuming that all of the initial kinetic energy of the car is absorbed by the universe as heat.

15.85

SET UP

In a poorly insulated bottle, water ($m_{\text{water}} = 0.350$ kg) and ice ($m_{\text{ice}} = 0.150$ kg) are initially at equilibrium at 273 K. Over time, the contents of the bottle come to thermal equilibrium with the outside air temperature, $T_{\text{air}} = 298$ K. Heat is transferred from the air to the ice and water—first, the ice melts and then all of the water heats up to 298 K. The heat necessary to melt the ice is related to its latent heat of fusion $\left(L_F = 334 \times 10^3 \frac{\text{J}}{\text{kg}}\right)$; the heat necessary to increase the temperature of the water is related to its specific heat $\left(C_{water} = 4186\dfrac{J}{kg \cdot K} \right)$. The

entropy changes of the melting ice and air are equal to $\Delta S_{melting} = \dfrac{Q_{melting}}{T_{ice}}$ and $\Delta S_{air} = \dfrac{Q_{air}}{T_{air}}$,

respectively. The temperature of the water is not constant, so we need to integrate

$dS_{warming} = \dfrac{mC_{water}dT}{T}$ in order to calculate the change in entropy of the water as it warms

from 273 K to 298 K. The total change in entropy of the universe is the sum of these three entropy changes.

SOLVE
Heat transferred:

$$Q_{water} = m_{water}C_{water}(T_{eq} - T_i) = (0.350 \text{ kg})\left(4186\frac{J}{kg \cdot K} \right)((298 \text{ K}) - (273 \text{ K})) = 36{,}627.5 \text{ J}$$

$$Q_{ice} = Q_{melting} + Q_{warming} = m_{ice}L_F + m_{ice}C_{water}(T_{eq} - T_i) = m_{ice}(L_F + C_{water}(T_{eq} - T_i))$$

$$= (0.150 \text{ kg})\left(\left(334 \times 10^3\frac{J}{K} \right) + \left(4186\frac{J}{kg \cdot K} \right)((298 \text{ K}) - (273 \text{ K})) \right) = 65{,}797.5 \text{ J}$$

$$Q_{air} = -(Q_{water} + Q_{ice}) = -((36{,}627.5 \text{ J}) + (65{,}797.5 \text{ J})) = -102{,}425 \text{ J}$$

Entropy change of the ice melting:

$$\Delta S_{melting} = \frac{Q_{melting}}{T_{ice}} = \frac{50{,}100 \text{ J}}{273 \text{ K}} = 183.5\frac{J}{K}$$

Entropy change of the water warming:

$$\Delta S_{warming} = \int_{273 \text{ K}}^{298 \text{ K}} \frac{dQ}{T} = \int_{273 \text{ K}}^{298 \text{ K}} \frac{mC_{water}dT}{T} = mC_{water}[\ln(T)]_{273 \text{ K}}^{298 \text{ K}}$$

$$= ((0.150 \text{ kg}) + (0.350 \text{ kg}))\left(4186\frac{J}{kg \cdot K} \right)\ln\left(\frac{298 \text{ K}}{273 \text{ K}} \right) = 183.39\frac{J}{K}$$

Entropy change of the air:

$$\Delta S_{air} = \frac{Q_{air}}{T_{air}} = \frac{-102{,}425 \text{ J}}{298 \text{ K}} = -343.7\frac{J}{K}$$

Entropy change of the universe:

$$\Delta S = \Delta S_{melting} + \Delta S_{warming} + \Delta S_{air} = \left(183.5\frac{J}{K} \right) + \left(183.4\frac{J}{K} \right) + \left(-343.7\frac{J}{K} \right) = \boxed{23.1\frac{J}{K}}$$

REFLECT

We can only use $\Delta S = \dfrac{Q}{T}$ if the temperature is constant. This formula follows directly from the

integral form: $\Delta S = \int ds = \int \dfrac{dQ}{T} = \dfrac{1}{T}\int dQ = \dfrac{Q}{T}$.

15.89

SET UP

A Carnot engine extracts heat Q_H from seawater ($T_H = 291$ K) to power a ship. The exhausted heat Q_C is used to sublimate a reserve of dry ice of mass m at a temperature $T_C = 195$ K. The ship's engines run at 8000 horsepower, where 1 hp = 746 W, for a full day; multiplying the engine's power by the time will give the work the engine is required to output to run the ship. We can use this value, the efficiency of a Carnot engine, and the fact that $W = Q_H - |Q_C|$ to find the amount of heat Q_C that will be exhausted. All of the exhausted heat will go to sublimating the dry ice, which means $Q_C = mL_F$; the latent heat of sublimation of carbon dioxide is $L_S = 573,700\dfrac{\text{J}}{\text{kg}}$.

SOLVE

Energy required for one day:

$$W = P\Delta t = \left(8000 \text{ hp} \times \frac{746 \text{ W}}{1 \text{ hp}}\right)\left(1 \text{ day} \times \frac{24 \text{ hr}}{1 \text{ day}} \times \frac{3600 \text{ s}}{1 \text{ hr}}\right) = 5.156 \times 10^{11} \text{ J}$$

Efficiency of the engine:

$$e = 1 - \frac{T_C}{T_H} = 1 - \frac{195 \text{ K}}{291 \text{ K}} = 0.3299$$

Total heat exhausted:

$$e = \frac{W}{Q_H} = \frac{W}{W + |Q_C|} = \frac{W}{W + |mL_S|}$$

$$m = \frac{W(1 - e)}{e|L_S|} = \frac{(5.156 \times 10^{11} \text{ J})(1 - 0.3299)}{(0.3299)\left(573,700\dfrac{\text{J}}{\text{kg}}\right)} = \boxed{1.83 \times 10^6 \text{ kg}}$$

REFLECT

An engine uses heat to do work; a refrigerator does work to remove heat.

Get Help: Picture It – Carnot Engine

Interactive Exercise – Leaky System

15.93

SET UP

When a 60-kg woman walks up five flights of stairs ($\Delta h = 20$ m), she normally releases 100 J of heat; if she has a high fever, she gives off 10% more heat. The work done by the person to climb the stairs is equal to the change in her gravitational potential energy. The efficiency of the woman's body is equal to the work done divided by the total energy she outputs, which is equal to the work done plus the heat emitted.

SOLVE

Work done:

$$W = \Delta U = mg\Delta h = (60 \text{ kg})\left(9.8\frac{\text{m}}{\text{s}^2}\right)(20 \text{ m}) = 11,760 \text{ J} = 11.76 \text{ kJ}$$

Normal temperature:

$$Q_C = 100 \text{ kJ}$$

$$Q_H = W + Q_C = (11.76 \text{ kJ}) + (100 \text{ kJ}) = 111.76 \text{ kJ}$$

$$e = \frac{W}{Q_H} = \frac{11.76 \text{ kJ}}{111.76 \text{ kJ}} = \boxed{0.105 = 10.5\%}$$

Elevated temperature:

$$Q_C = 1.1(100 \text{ kJ}) = 110 \text{ kJ}$$

$$Q_H = W + Q_C = (11.76 \text{ kJ}) + (120 \text{ kJ}) = 121.76 \text{ kJ}$$

$$e = \frac{W}{Q_H} = \frac{11.76 \text{ kJ}}{121.76 \text{ kJ}} = \boxed{0.0966 = 9.66\%}$$

Your $\boxed{\text{efficiency drops when you have a fever}}$.

REFLECT
Since her efficiency is lower when she is sick, it will take her more energy to accomplish the same tasks she did when healthy.

15.97

SET UP
The temperature increases by 3.0 degrees C for every 100 m we drill into the Earth's crust. Water pumped into an oil well that is 1830 m deep is used as a heat engine. The temperature of the hot reservoir (that is, the bottom of the well) is equal to the depth multiplied by the relationship between the temperature increase and depth. The temperature of the cold reservoir (that is, the surface) is $T_C = 293$ K. The maximum efficiency occurs when the heat engine acts reversibly, so $e = 1 - \dfrac{T_C}{T_H}$. We want to use a combination of these wells to create a reversible power plant that provides 2.5×10^6 W of power. We can use a different expression for efficiency, $Q_H = \dfrac{W}{e}$, in order to calculate the total heat the power plant absorbs from the Earth in one day.

SOLVE

Part a)

Maximum temperature:

$$T_H = (1830 \text{ m})\left(\frac{3.0°C}{100 \text{ m}}\right) = 54.9°C - 328 \text{ K}$$

Maximum efficiency:

$$e = 1 - \frac{T_C}{T_H} = 1 - \frac{293 \text{ K}}{328 \text{ K}} = 0.107 = \boxed{11\%}$$

Part b)

$$e = \frac{W}{Q_H}$$

$$Q_H = \frac{W}{e} = \frac{(2.5 \times 10^6 \text{ W})\left(1 \text{ day} \times \frac{24 \text{ hr}}{1 \text{ day}} \times \frac{3600 \text{ s}}{1 \text{ hr}}\right)}{0.107} = \boxed{2.0 \times 10^{12} \text{ J} = 2.0 \text{ TJ}}$$

REFLECT

A terajoule (TJ) is equal to 1×10^{12} J. Because we used the maximum efficiency, 2.0 TJ is the minimum possible energy that is absorbed from the interior of the Earth in one day. If the actual efficiency is smaller than 11%, the power plant must absorb more than 2.0 TJ of heat in a day in order to deliver a power of 2.5 MW.

Get Help: Picture It – Carnot Engine
Interactive Exercise – Leaky System
P'Cast 15.9 – Actual and Theoretical Efficiencies

15.101

SET UP

A heat engine absorbs $Q_H = 1250$ J from a high-temperature reservoir ($T_H = 490$ K) and does 475 J of work. The temperature of the cold reservoir is $T_C = 273$ K. The efficiency of the engine is equal to the work done divided by the heat absorbed. The change in the entropy of the universe is equal to the sum of the entropy changes of the hot and cold reservoirs. The heat exhausted to the cold reservoir Q_C was not given but is equal to $Q_H - W$ from the first law of thermodynamics. Finally, the amount of energy unavailable for doing work after one full cycle is equal to $T_C \Delta S$.

SOLVE

Part a)

$$e = \frac{W}{Q_H} = \frac{475 \text{ J}}{1250 \text{ J}} = \boxed{0.38 = 38\%}$$

Part b)

$$\Delta S = \Delta S_H + \Delta S_C = \frac{Q_H}{T_H} + \frac{Q_C}{T_C} = \frac{Q_H}{T_H} + \frac{Q_H - W}{T_C}$$

$$= \frac{-1250 \text{ J}}{490 \text{ K}} + \frac{(1250 \text{ J}) - (475 \text{ J})}{273 \text{ K}} = \boxed{0.288 \frac{\text{J}}{\text{K}}}$$

Part c)

$$W_{\text{unavailable}} = T_C \Delta S = (273 \text{ K})\left(0.288 \frac{\text{J}}{\text{K}}\right) = \boxed{78.6 \text{ J}}$$

REFLECT

The amount of energy unavailable to do work is also equal to the difference between the maximum work possible in a reversible process and the actual work done by a system.

15.105

SET UP

A certain person typically eats about 2250 kcal per day, 80% of which is transferred to the surroundings as heat. The surroundings are at 295 K. The entropy change of the surroundings is equal to the heat absorbed divided by the temperature.

SOLVE

$$\Delta S = \frac{Q}{T} = \frac{(0.80)\left(2250 \times 10^3 \text{ cal} \times \dfrac{4.184 \text{ J}}{1 \text{ cal}}\right)}{295 \text{ K}} = \boxed{2.55 \times 10^4 \frac{\text{J}}{\text{K}} = 25.5 \frac{\text{kJ}}{\text{K}}}$$

The entropy of the apartment $\boxed{\text{increases}}$.

REFLECT

The surroundings absorb heat, so the entropy of the surroundings should increase.